Carl Manchot
The Cutaneous Arteries of the Human Body

Foreword by G. Ian Taylor
Introduction by William D. Morain

Translated by Jovanka Ristic
and William D. Morain

With 53 Figures and 8 Color Plates

Springer-Verlag
New York Berlin Heidelberg Tokyo

William D. Morain, M.D.
Associate Professor of Surgery (Plastic)
Dartmouth-Hitchcock Medical Center
Hanover, New Hampshire 03755, U.S.A.

Jovanka Ristic, M.S.
Reference Librarian
American Geographical Society Collection
University of Wisconsin/Milwaukee Library
Milwaukee, Wisconsin 53211, U.S.A.

Sponsoring Editor: Marie Low
Production: Anthony Buatti
Design: Caliber Design Planning, Inc.

Library of Congress Cataloging in Publication Data
Manchot, Carl.
   The cutaneous arteries of the human body,
   Translation of: Hautarterien des menschlichen Körpers.
   Bibliography: p.
   1. Skin—Blood-vessels.   2. Arteries.
3. Anatomy, Human.   I. Title.   [DNLM: 1. Skin
—Blood supply.   2. Arteries.   WR 101 M268h]
QM481.M3513   1983   611'.77   83–4704

Typeset by Kingsport Press, Kingsport, Tennessee.

9 8 7 6 5 4 3 2 1

ISBN-13: 978-1-4613-8223-2   e-ISBN-13: 978-1-4613-8221-8
DOI: 10/1007/978-1-4613-8221-8

To Gertrud Manchot

Contents

Foreword

Sir Harold Gillies often said: "Plastic surgery is a constant battle between blood supply and beauty." While he championed the cause of the poor unfortunates of the two Great Wars, he laid down his principles for tissue transfer. These dicta were based, like warfare, on attack and counterattack. With the hand of an artist he molded and advanced the tissues of the body with geometrical precision to their limits of vascular endurance, only to be rebuffed on occasions by necrosis. Gradually, however, with a dynamic army of colleagues and a meticulous intelligence corps who recorded every strategic move, he won new ground and created an empire that evolved into Plastic Surgery as we know it today.

Among his 16 commandments was the one that stated, "Consult other Specialists." Interestingly, this was the thirteenth dictum and perhaps should have read, "Consult other Specialists and translate their works." Gillies did not speak German and craftily concealed this fact when, assembled before a German audience in 1939, he gesticulated to the assembled multitude while a gramaphone, cunningly hidden in the podium, delivered his lecture. This language barrier was unfortunate, for half a century earlier Manchot had published his treatise "Die Hautarterien des Menschlichen Körpers"—the skin arteries of the body. As an anatomist Manchot laid the foundation for the revolution in skin flap design that has erupted in the last two decades and has taken vascularized tissue transfer from a geometrical trial to an anatomical and a biological certainty.

With the enthusiasm and energy of youth, Manchot performed an exhaustive anatomical study of the cutaneous vessels of the body, completing this treatise in the incredible time of 6 months. His anatomical drawings, which locate the origins of these vessels from the musculo-aponeurotic planes, were executed with great accuracy and in the main have stood the test of time. His scientific method, however, has been lost in the mists of antiquity. Many attempts have been made to unravel this mystery, including those of William Morain. In the preface that follows, Dr. Morain describes in graphic detail his adventure into the archives of Hamburg in search of the answer. A chance comment from one of his peers, a book on a dusty shelf in a library, a fortuitous meeting with a German colleague, a name in a telephone directory, and the hunt was on. Just when his prey was in sight—frustration! There in the bowels of the Institut d'Anatomie Pathologique de Strasbourg was a jumbled pile of unlabelled specimens—some Carl Manchot's, the others the efforts of the forgotten.

Whatever his technique, Manchot almost certainly injected a substance into the arterial system of his cadaver specimens, which he then dissected. His illustrations attest to this. They show the cutaneous arteries and their main diversions but do not show the finer ramifications and *arterial* connections that we have found to exist between all adjacent vascular territories. Unfortunately, Roentgen had yet to make his discovery when Manchot published his work, and it was not until nearly a half-century later that Michael Salmon demonstrated these connections. Salmon reappraised the work of Manchot, but he had the advantage of X-rays. Unfortunately, he too was on the opposite side of the channel from Gillies and spoke a different tongue. As with Manchot, his work lay dormant for decades and has been only recently discovered in the English-speaking medical world. A letter from Dr. B. G. H. Lamberty and a chance comment from an aspiring young plastic surgeon, Dr. Rosa Razaboni, drew my attention in only recent months to the work of Salmon. "I believe I have a book that may interest you. It was a gift from the late Dr. Converse." There in a disintegrating volume of "Arteres de la Peau" was the missing link and the confirmation of the results of our investigations into the vascular territories of the body. William Morain has kindly asked me to give an account of this research, but first the story must be completed.

At the turn of the century advances on the clinical front gave new significance to the work of Manchot. Tansini reported a latissimus dorsi flap in 1906, and in 1918 and 1931, Esser charted various flaps in the head and neck based on named arteriovenous systems. In North America events were happening as well. In 1921 Blair described the forehead flap based on the superficial temporal vessels, although he advised transferring this flap after a delay. In 1939 Webster examined the work of Manchot and described a long bipedicled thoraco-"epigastric" flap, based on named arteriovenous systems, which extended from the groin to the axilla. Shaw in wartime practice utilized each end of these systems separately to provide one-stage direct flaps for hand reconstruction.

It is with the benefit of hindsight that we observe the lack of communication during this period, not only between anatomists and surgeons but between surgeons themselves. Perhaps it was distance, the language barrier, the special interests, the personalities involved, or just the overwhelming volume of day to day clinical problems that prevented the collaboration of ideas. What would Plastic Surgery have been a half-century ago if a summit meeting had been possible between Manchot, Salmon, Gillies, Blair, Tansini, Esser, and the other great names of the era.

Bakamjian drew attention to the long paramedial perforators of the internal mammary system in 1965. Then a giant step forward was taken in 1972 when McGregor and Jackson cemented the efforts of the anatomists and the surgeons by differentiating between large flaps based on a known axial blood supply and those relying on random vessels in the area. This culminated in their discovery of the groin flap. Daniel and Williams reappraised the work of Manchot and others, and on an embryological and an anatomical basis segregated the cutaneous arteries into direct cutaneous and musculo-cutaneous.

When "The Free Flap" was published in 1973, surgeons scurried back to the dissection room to find other donor sites for such transfer. At first the known axial skin flaps were utilized, and almost every square inch of the integument was assayed for possible dispatch to a distant site.

In the same year that the free flap was launched, McCraw and coworkers from Atlanta and Norfolk surfaced with their submarine of skin and muscle that announced the revival of the myocutaneous flap. Remarkably, McCraw and Vasconez had set out initially to complete the work of Ger and chart the vascular anatomy of the human muscles to be used for microvascular free transfer. They deduced that their experimental island muscle flaps would "carry overlying skin." By a quirk of fate it was realized that this unit had more potential as a local flap, and so McCraw and Dibbell emerged from the experimental laboratory into the clinical arena with the island biceps femoris myocutaneous flap—nearly 70 years after Tansini's report of the first latissimus dorsi flap.

Rapidly the body became a checkered map of donor sites for myocutaneous flaps, and the race was on. As fast as a free flap was used for a different recipient site, it was torpedoed by a flurry of muscle or myocutaneous flaps that emerged in the local area. Many of these techniques provided aesthetic and functional results, while others were, as the late Frank McDowell has described, "globs and blobs." This led to the era of the fasciocutaneous flap, a period that is still in a state of flux. To escape "the hamburger" of muscle and skin, workers looked for another solution and soon realized that blood vessels follow fascial planes. Donor sites have since emerged that include the deep fascia and the intermuscular septa.

Well, where is this all leading? Random flaps, axial flaps, myocutaneous flaps, fasciocutaneous flaps—are these all separate entities, or are they a different clinical expression of the same cutaneous architecture that Manchot had told us about nearly a century ago? If an understanding of the fundamentals of embryonic tissue differentiation and migration were combined with a knowledge of the works of Manchot and Salmon, then a visit to the Hunterian museum at The Royal College of Surgeons in London would answer this and other questions regarding the cutaneous circulation. There, a series of magnificent specimens display the entire arterial "skeleton" of the body—yet another labor of love by anatomists of a different era.

Our own investigations of the cutaneous vascular architecture began a decade ago when we dissected fresh cadavers for free flap donor sites. Later we injected the supplying arteries alternatively with red and black india ink to define their cutaneous distribution. This technique was validated experimentally in the pig. Recently we have investigated the origin, course distribution, and connections of these vessels by injecting various radio-opaque substances into the arterial system. The radiographic studies were commenced 18 months before we became aware of the work of Salmon. Remarkably, our technique was almost identical to his, but with some variations. These vessels were accurately located by first obtaining radiographic studies of the region and then subtracting the integument from the underlying musculoskeletal tissues and X-raying them separately. A full account of this work is not possible nor appropriate in this context. Some investigations are incomplete, and some completed works have yet to be published. However, the following is a chronological account of the important investigations, observations, and conclusions.

Initial fresh cadaver dissections revealed two important facts. The first was the arterial connections that exist between vessels of adjacent territories, and this was well demonstrated in the groin region. The second observation was the interrelationship between the size of vessels in adjacent territories in different individuals; when the territory of one artery was large, the territory of its companion was often small. Typically this interrelationship was seen in the deltopectoral flap between the internal mammary perforators and the cutaneous branches of the acromio-thoracic axis and in the groin region between the superficial circumflex iliac and the superficial inferior epigastric vessels.

When ink was injected into a cutaneous artery, it produced skin blotches corresponding to the major cutaneous branches. Further injection resulted in smaller patches of staining between and beyond the initial area of dye, until finally these areas began to coalesce and stain more deeply as the dermal circulation was filled. A stage was finally reached where further injection did not extend the perimeter of skin coloration, although occasionally it appeared in blotches at a remote site. *In no case, where the vessel of a known flap was injected in the intact body, did the perimeter of cutaneous skin staining extend beyond the expected area of survival of such a potential flap.* In most cases it fell well short of that margin.

When large volumes of ink were used, often the dye was seen to appear from the cannula inserted into an adjacent artery or to escape into the general circulation and appear at a remote site such as the viscera. The reason for this "overflow phenomenon" is revealed in our radiographic studies (see Figure C). They illustrate the reduced caliber *arterial* connections that link adjacent vascular territories.

In the fresh cadaver the ink follows the line of least resistance. When dye is injected into a cutaneous artery, some is distributed via its branches to the skin. The remainder passes via these "choke" vessels into the branches of the adjacent system. Further dye is again distributed to the skin until it reaches the main stem of the adjoining arteries. It then escapes in retrograde fashion into the general circulation rather than perfusing a succession of linked cutaneous arteries of small calibre.

In this way the vascular territories of the various cutaneous arteries were mapped on the body with red and black ink (Figure A). Considerable overlap in skin staining was found between adjacent territories in contrast to the circumscribed areas charted by Manchot (Figure B). This overlap would be even greater if allowance were made for the unstained area of skin that would survive if a flap were based on any particular vessel.

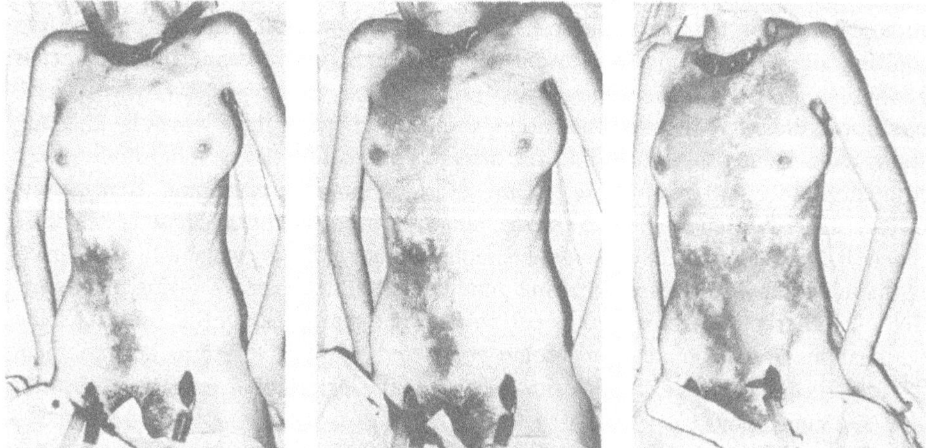

Figure A.   Progressive injection studies of the deep inferior epigastric artery (*left*), the internal mammary–superior epigastric system (*center*), and the deep circumflex iliac artery of each side, which overlap and stain most of the anterior torso (*right*). Reprinted courtesy of Plastic and Reconstructive Surgery, Vol. 72, 1983 (in press).

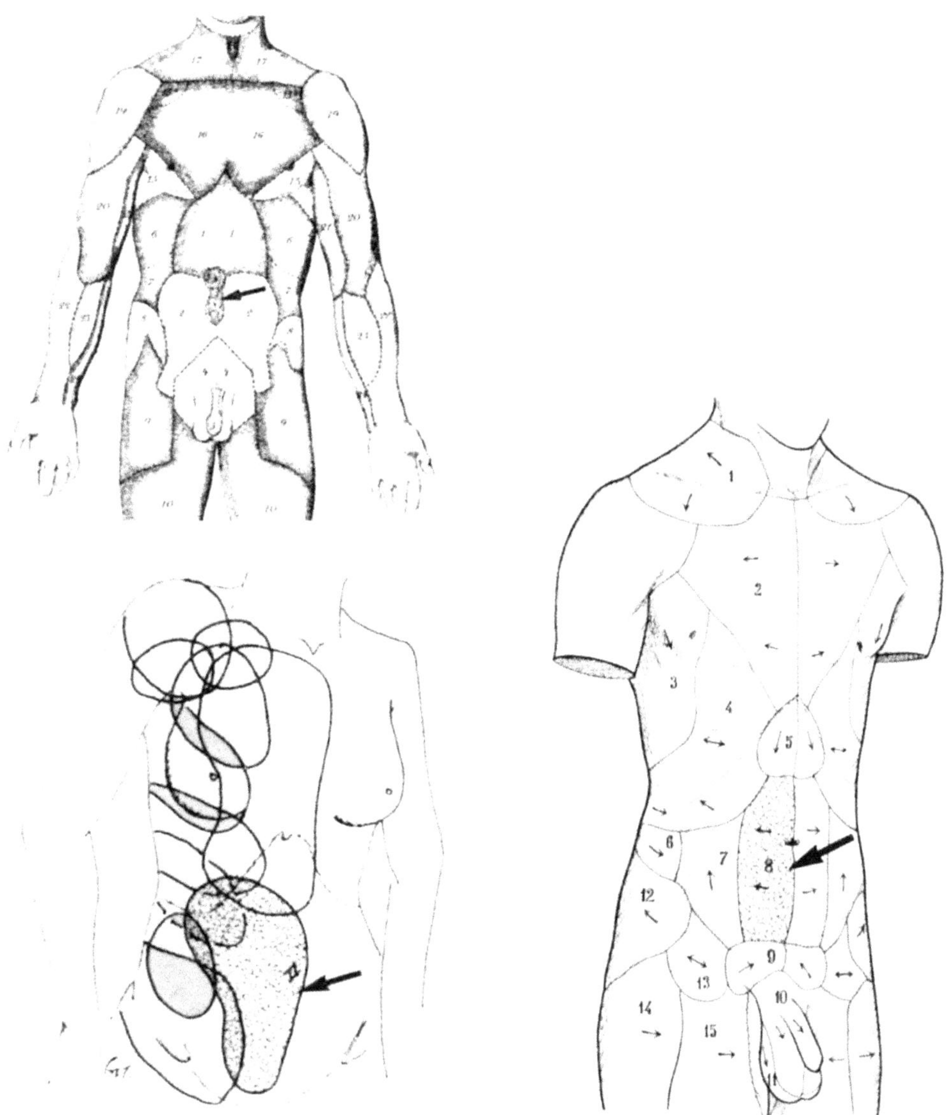

Figure B. The "anatomical" territories of Manchot (*top left*) and Salmon (*right*), compared with our "potential clinical" territories (*bottom left*) on the torso. Note the considerable overlap of these vascular territories in the latter diagram, which are depicted from the top down as the four branches of the acromio-thoracic axis, the internal mammary superior epigastric system (unshaded), the lateral thoracic and thoraco-dorsal system, just two of the lateral intercostal vessels; the deep inferior epigastric artery (*arrow*), the deep circumflex iliac artery, and finally its superficial counterpart. The remaining intercostal, lumbar, superficial inferior epigastric, and pudendal vessels have been omitted for clarity. In each illustration the territory of the deep inferior epigastric artery has been shaded and marked with an arrow for comparison. Salmon's territories were designated as follows: 1: Artères cervicale transverse et sus-scapulaire; 2 and 5: A. mammaire interne (2, branches directes, 5, A. épigastrique superficielle supérieure); 3: A. mammaire externe et scapulaire inférieure; 4: Rameaux perforants des artères intercostales; 6: Rameaux perforants des artères lombaires; 7: A. épigastrique superficielle inférieure; 8: A. épigastrique profonde; 9: A. honteuse externe supérieure; 10: A. honteuse externe inférieure; 11: A. honteuse interne; 12: A. circonflexe iliaque superficielle; 13: A. fémorale primitive, branches directes; 14: Grande artère du vaste externe; 15: A. fémorale superficielle.

The reason for this disparity is evident. Manchot's mapped areas depict the *anatomical* territory of the cutaneous arteries and their branches. They are based on cadaver dissections, limited by the naked eye. Salmon has given a more precise map of these anatomical territories, which nevertheless approximate in many areas to those of Manchot (Figure B). His radiographic studies take into account the finer ramifications of these arteries and their connections with adjacent vessels. Our dye injections go beyond the anatomical territory of the injected artery and stain the skin supply of the primary connections of this vessel with adjacent cutaneous arteries. Therefore these injection studies depict the *potential clinical* territory of the artery. However, since dye escapes retrogradely into the general circulation via the main stems of the adjacent vessels, it represents a conservative estimate of the total area of integument that would survive as a flap if based on the injected artery.

Dissection of the deeper tissues revealed that the predominant supply to the skin followed fascial planes either *between* muscles in the *intermuscular* septa or between the muscular bundles in the *intramuscular* septa as *fasciocutaneous* vessels. These vessels

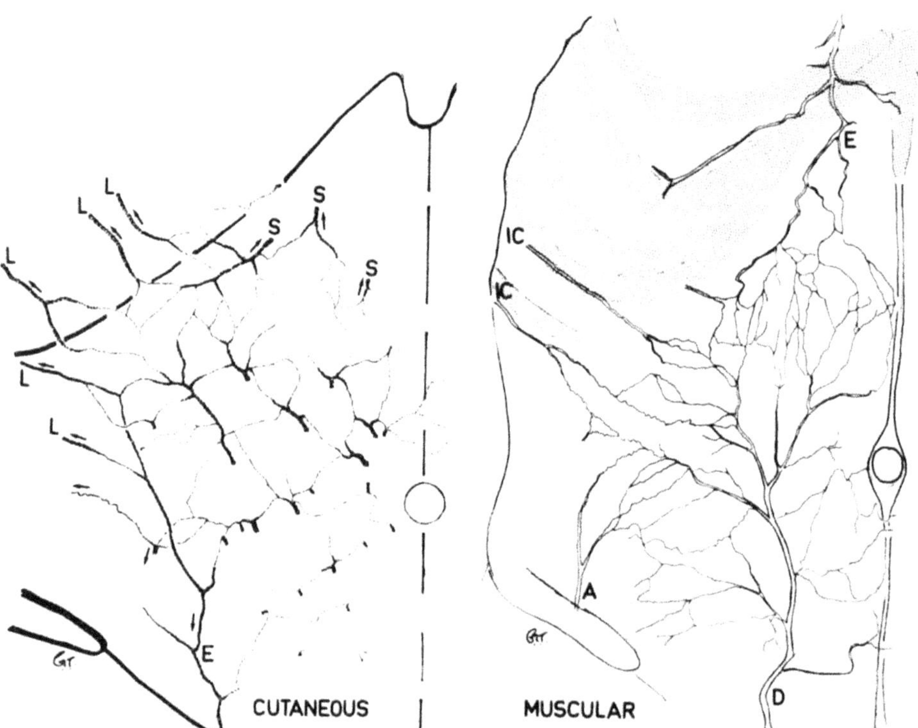

Figure C. Tracing of barium radiographic studies of the deep inferior epigastric artery (D) that reveals the mirror image of its branches and their connections in the integument (*left*) and in the underlying musculo-skeletal layer (*right*). In both layers these branches radiate from the fixed paraumbilical hub in "spoke wheel" fashion. In the muscular layer they communicate via "choke" arteries, with branches of the superior epigastric artery (E), the intercostal arteries (IC), and the ascending branch of the deep circumflex iliac (A). In the integument the cutaneous perforators of the deep inferior epigastric artery link with the similar branches of the superior epigastric artery (S), the lateral intercostal perforators (L), and the superficial inferior epigastric artery (E). Reprinted courtesy of Plastic and Reconstructive Surgery, Vol. 72, 1983 (in press).

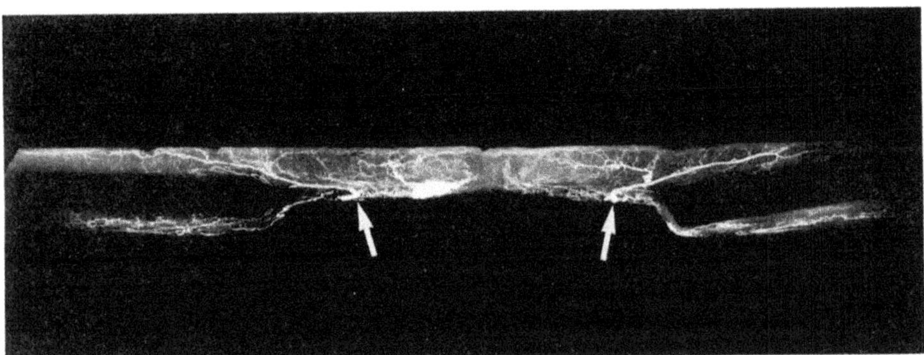

Figure D. Transverse section of the ventral abdominal wall at the umbilical level in a fresh cadaver after barium injection of the deep inferior epigastric systems. The muscle layer has been dissected from the integument to the lateral border of the rectus muscle. Note the orientation and connections of these vessels in the peritoneum, the muscle, on the undersurface and within the subcutaneous fat, and in the subdermal plexus. Note also the course of the large musculocutaneous perforators of the deep inferior epigastric artery (*arrows*) on each side as they pass from the fixed skin area over the rectus sheath to reach the mobile skin area above the external oblique aponeurosis. Reprinted courtesy of Plastic and Reconstructive Surgery, Vol. 72, 1983 (in press).

are well illustrated by Manchot in the latter part of this book. They are particularly well developed in the following sites: adjacent to the midventral and middorsal line of the torso, in the axillary line, over the flexor surface of joints, and radiating up from the skull base, where they are known as direct cutaneous arteries. They are also concentrated in the intramuscular septa of the extremities, where they arise from nearby segmental arteries or their main branches.

After piercing the deep fascia, the vessels of adjacent territories form a rich anastomosis on its superficial surface. They penetrate the subcutaneous fat between the fat locules and follow connective tissue septa to the subdermal plexus of the skin. In general the dominant cutaneous vessels emerge from the musculo-aponeurotic plane in fixed skin areas of the body and radiate to mobile skin areas—the greater the mobility the longer the vessels. The lines of attachment of the integument to the underlying tissue are easily seen in a thin muscular individual at flexural skin creases, in the midline of the torso, and over the intermuscular and intramuscular septa. They mark the site of origin of these cutaneous vessels from the deep tissues.

The dye injection studies have provided the experimental basis for composite tissue transfers. Fresh cadaver dissections often revealed contributions from the injected artery to the underlying soft tissues and bone. Communications between vessels of adjacent territories were again seen in all layers. Consequently, various combinations of skin, tendon, muscle, nerve, and bone have been transferred successfully en bloc to a distant site by microvascular techniques, nourished by a single arteriovenous system.

Our radiographic studies have confirmed many of the above findings. They support the work of Salmon and reveal the cutaneous circulation to be a continuous network of linked vascular territories. The connections consist of reduced caliber "choke" vessels that generally consist of arteries of less than 0.5 mm diameter but can be seen in the smallest radicles of the arterial tree (Figures C and D). They exist in all layers between the skin and the deep fascia and follow connective tissue planes and septa.

When the radiographic substance, usually barium, was injected into an artery, it was seen to progressively "borrow" and infiltrate in chain link fashion the adjacent arterial network.

The orientation of the cutaneous vessels and their interconnections were often seen mirrored by their parent segmental arteries as these vessels coursed in the underlying musculoskeletal plane. This interesting relationship was revealed by our subtraction technique and is illustrated in Figures C and D, where the deep inferior epigastric artery has been examined. The same vascular pattern was reflected in the peritoneum.

The above snippets of information, when combined with the giant efforts of Manchot and the subsequent contributions of Salmon, would suggest that *the blood supply of the body is a reflection of its connective tissue superstructure.* The life of Carl Manchot and his account of this vascular framework to the skin is revealed for us in English thanks to the diligence and eloquence of Dr. Morain and Springer-Verlag. Dr. Morain's labors have been justifiably honored, and this volume demands a place on every plastic surgeon's bookshelf.

Melbourne, 1983

G. Ian Taylor, F.R.A.C.S., F.R.C.S.
Hunterian Professor

Preface

This book has sprung from the fused chrysalis of three families: plastic surgery's, Manchot's, and my own. That the significance of the nineteenth century monograph and its author, lying dormant for so many years, should emerge to give wings to a proud surgical specialty is the message I have wished to convey. Gathering the pieces of this story has been a family affair from the beginning.

I do not speak much German. If I did, I would probably have been content with the tea and conversation offered so freely by my wife Dagmar's articulate 89-year-old grandmother under the restful castle near the Dutch border where we often vacationed. In the spring of 1979, on my fourth legitimate attempt to prepare for a trip, I finally admitted linguistic defeat amidst a cataract of personal pronouns on page 97 of *German for Beginners*. When I explained that I wanted to travel for some historical research during our approaching summer trip, my wife seemed actually relieved at the prospect of having other than a smiling, indolent mute to display to her family again that year. So it was agreed.

Dr. Ernest Kaplan had introduced me to the name Manchot at Stanford in 1974. We were having a hurried conversation over the new axial flaps as we entered the large secretarial suite: "But what *is* the blood supply to the skin?" I asked. "Oh, we've got that. Some German did it in the 1800s. Name was Manchot. We've got a Xerox copy from the National Library of Medicine in Washington." He was gone.

I remember standing alone and laughing as he disappeared into his inner office. The impact of his words was immediate: all those years of ignoring the blood supply to our skin flaps and it had been available all along. But who was that anatomist again?

For some reason the conversation had come back to me on page 97 of *German for Beginners*, and my way out was clear. If I could not adapt graciously to the *Konditorei* lifestyle, Germany's unknown anatomist would at least be found. Mohammed would go to the mountain.

Dagmar produced the catalyst from her family. A cousin, Detlev Feldkamp, was in his first year of medical studies at the nearby University of Münster. He had earlier spent a year in the Caribbean on an exchange program and spoke superb English. He agreed to lend his help. Armed with Detlev's library card and the name "C. Manchot," we assaulted the university library one afternoon. We did not learn much. The book itself was on the open stacks. We checked it out, photocopied it, and sat down to translate the "Foreword." Suspecting the influence of a Professor Dr. Schwalbe mentioned in the two paragraphs, we retreated to the card catalog and confirmed a contemporary Professor Gustav Schwalbe in Strassburg. Following a hunch, we checked out his major lifework on the anatomy of the sensory organs and found our first bona fide lead in his short chapter on cutaneous innervation.

We were less successful with Manchot himself. No reference could be found to the "Professor" in any *Who's Who* or other biographical guide. How could such a great anatomist have disappeared?

Our search of the German counterpart of *Index Medicus* gave a clue. There were four or five subsequent papers cited under the name "Manchot, C", extending all the way to 1930. One of them, dated 1903, was entitled (in translation) "Delirium Tremens and the Question of Certificate of Need in the Granting of Liquor Licenses in Hamburg." It was a long shot, but we had a city.

On our return to the States, I had some exploratory and generally discouraging correspondence with several publishers about the project. One who was more favorably inclined, Robert Krieger of New York, offered the name of Heinz Graupe, a publishing consultant and scholar of Jewish intellectual history who happened to live in Hamburg. Mr. Krieger suggested that Herr Graupe might be able to provide personal assistance should we wish to travel to Hamburg in our searches on a second visit abroad.

I wrote to Graupe, briefly outlining the project and our tenuous lead concerning Hamburg and leaving an open-ended appeal for suggestions. His reply arrived in English within the month. "Dear Dr. Morain," he wrote, "After receiving your very interesting letter from June 16th I found in the telephone directory twice the rather unusual name Manchot. The first name I contacted turned out to my pleasure to be a daughter of Dr. Manchot: Miss Gertrud Manchot, a tax advisor and chartered accountant." A ten-strike!

My letter to Frau Manchot produced a wealth of information in return: a three-page letter brimming with details of her father's life, a couple of lengthy obituaries, and an invitation to visit her in Hamburg. She possessed a copy of the book, but knew little of its origins.

Since it was now clear that the story could be developed, a decision had to be made concerning a full translation of the original monograph. At least one such translation had been done previously, and it was to be over a year before I could determine with certainty that no published English version had reached print. Thus, it was a gamble when I asked Dagmar's sister, Jovanka Ristic, of Shorewood, Wisconsin, to draft a translation of Manchot's work. As each perfectly typed installment arrived by mail, it remained only to choose among the carefully selected alternative phrases she provided, to translate Manchot's Latin terms, and to make adjustments for contemporary preferred terminology. This did not completely avoid problems, however, as in the case of one forgettable 2½-hour evening of frustration over an irreconcilable discrepancy between text and diagrams in the chapter on the upper extremity. This was resolved only with the belated realization that the word "anconeus" referred to the *triceps* muscle and not to what we presently understand as the anconeus.

The translation, itself, is close to a literal one. We have preserved Manchot's original paragraph delimitations, even when they seemed inappropriate. To have varied liberally from the flow of the original seemed to have been too much license.

As the next summer approached, bringing with it elaborate family plans for a 90th birthday party for Oma, Detlev and I were making our own date in Hamburg. But one facet of the story was still uninvestigated. I had written months before to the Université Louis Pasteur in Strasbourg, France, in an attempt to establish a useful contact for researching the original Manchot investigations. A short letter in April from the Director of the Institut d'Anatomie Normale assured me there was no useful information to be found concerning such a dated book. It was unfortunately not until returning from abroad that I found a second letter waiting on my desk that would provide a very different answer.

Our visit with the gracious Gertrud Manchot was even more than we had anticipated. Not only was she the photo collector and chronicler of the family, but she had coincidentally assumed the personal effects of an aging physician sister several weeks before. In those effects were some of the original manuscript, sketches, and illustrations. We spent much of the afternoon in a taxicab, photographing the personal Manchot landmarks in the city. The rest of the time was spent carefully filling in details of the story (Figure E). Detlev proved to be a masterful interviewer and translator, and Frau Manchot clearly evinced the personal wisdom, spirit, grace, and pride that had so characterized her father.

The letter waiting on my desk was from Le Docteur Theodore Vetter, an urbane archivist with the Institut d'Anatomie Pathologique de Strasbourg. He had spent several days digging through dusty logbooks to produce a wealth of photocopies of letters, records, rosters, and other archival materials directly bearing on Manchot and his work. It was central to our researches and was to provide most of the answers to

Figure E.   Gertrud Manchot (left) and Detlev Feldkamp with the original manuscript.

the questions still lingering from the brief Foreword. This time Detlev took the principal role, taking with him to Strasbourg a young neighbor from Münster who spoke fluent French to assist in the meeting with Vetter. Detlev proved to be as good at photography as he was at English translation. It was a very productive trip, in large measure because of the interest of the brilliant Vetter and the eager enthusiasm of his interviewers.

There is still a critical point in the story waiting to be discovered in Strasbourg. It waits in the third floor attic storeroom. What was the material Manchot used for his injection studies? Vetter assured Detlev the dissections were still there and that the answer would be immediately evident on inspection. The only catch was that all of those specimens, along with hundreds of other important anatomical studies, had had to be protected from the bombs of World War II and were, therefore, hastily transported to the cellar *sans* labels. They were brought back to the third floor without sorting after the danger had passed. A single look at the gargantuan pile of dismembered parts was sufficient to remind Detlev that some remaining mystery is good for any story.

In addition to those I have already mentioned, the number of unselfish people who have contributed information, suggestions, and cherished criticism to the development of the final historical introduction is very substantial. I first of all wish to acknowledge the warm support and sense of mission conveyed to me from members of the Manchot family in both Germanys. I am particularly indebted to Dr. John McCraw for his many hours of valuable blue-penciling to smooth out the prose and to correct the modern historical record. To Dr. Paul Natvig and his brilliant colleague Dr. Renee Lang who provided needed cross-cultural insight and offered many valuable literary tips I am equally grateful. Dr. M. J. Jurkiewicz, Dr. Robert Mills, and Dr. John Head provided valuable perspectives in balance and historical fact. I thank Dr. Brian Ross, Dr. George Cherry, and Dr. Bert Myers for their recollections of Stuart Milton. To my newly found friends in the world of publishing—Neale Watson, Robert Krieger, Leslie Adams, Sherwin Nuland, and especially Robert Rowan—I gratefully acknowledge assistance in my odyssey to Springer-Verlag. A particular debt of gratitude is owed to Dr. Mark Gorney, who

Figure F.   Carl Manchot's grave in Hamburg.

took great personal interest in seeing that the material was presented publicly in the broadest forum. I thank Dr. Stephan Ariyan and Dr. Ulrich Weil of Yale who provided the final vital link to publication. Mr. Richard Wolfe, rare books librarian at Harvard's Countway Medical Library, was most helpful in many phases of my research and was kind enough to lend the fine Harvard copy for reproduction of the plates. I am also forever in debt to Mr. Wolfe for caroming an available original copy—one of half a dozen or so known to exist in the United States—from its discovery in an old New England farmhouse to my attention and ultimate acquisition. It was especially humbling to have been honored by Dr. Bard Cosman and the Webster Society with their 1981–82 award for a paper on plastic surgery history. I also wish to acknowledge my most loyal Marge Thibodeau and Joy Williams for their customary secretarial excellence. And finally, to Marie Low and Springer-Verlag I owe the greatest debt for having the confidence to invest their individual and corporate resources in this unusual project. It is perhaps the final irony that they became involved in the project during the week of the golden anniversary of Carl Manchot's death (Figure F).

Of course, like most of the rest of the good things in my life, none of this would have happened at all without the continuing participation—nay, the very existence—of Dagmar.

Hanover, New Hampshire
August, 1982

WILLIAM D. MORAIN, M.D.

Introduction

Carl Manchot: Plastic Surgery's Missed Opportunity

William D. Morain, M.D.

— I —

About 200 B.C. the Greek mathematician and cartographer Eratosthenes postulated that the world was round and set about establishing his theory by a grand Euclidean proof. He had observed that there was a deep well at Syene (now Aswan) in the upper waters of the Nile in which the sun's reflection could be seen only at noon on June 21. At Alexandria, 500 miles to the north, where he was curator of the large Hellenic library, he decided one year to conduct an experiment on the solstice. At midday he carefully measured the length of the shadow cast by the great lighthouse of Pharos, one of the Seven Wonders of the World and the tallest structure the Greeks had ever built. Knowing its height, this horizontal measurement and a simple calculation gave him a 7° solar intercept angle, $\frac{1}{50}$th of a circle. Multiplying by the 500 miles to Syene, he obtained the correct answer, a 25,000 mile circumference, clearly a brilliant achievement to be drawn from a single linear measurement.

Nearly two millennia later Christopher Columbus also postulated that the world was round and that one could therefore sail west to the Indies, but took instead a more empiric, perhaps surgical, attitude in determining his proof. Rather than first studying the Classics, he tacked his three ships westward, discovered land, called it the Indies, and pronounced its inhabitants to be Indians. Columbus went to his grave, however, wondering where the silk and spices were, having established a yet unbroken record of a 12,000-mile navigational error, the greatest in nautical history. It is said by some historians that Columbus died in prison; if so, it is perhaps a fitting end for someone of his particular origin who failed to heed the lessons of the Classics.

In 1889, while still a 23-year-old medical student in Strassburg, Carl Manchot published his own classic work, *Die Hautarterien des Menschlichen Körpers* (Figure 1). In careful detail he mapped the topography of the cutaneous circulation of the

Figure 1. Carl Manchot as a young medical student (courtesy
of Gertrud Manchot).

entire human body, drawn from a series of the most difficult and precise dissections
and injection studies. Surely such a resource would be eagerly sought out by those
surgical pioneers who would wish to develop the means and designs for transferring
blocks of skin from place to place!

As history would have it, precisely a quarter century after Manchot's publication,
the greatest flood of surviving human traumatic deformities in the history of warfare
began to pour out of both sides of the trenches of France and Belgium (Figure 2).
Waiting across the English Channel was Captain Harold Gillies, the young New Zealand
otolaryngologist who, with a few colleagues, established the first plastic surgical unit in
the world at Aldershot and then the historic 600-bed Queen's Hospital reconstructive
unit at Sidcup (Figure 3). Dr. Gillies, through the fiery baptism of the Great War,
truly forged the modern specialty of plastic surgery, establishing the fundamental principles of tissue transfer and reconstruction in empirical fashion as the day-to-day contingen-
cies demanded. As David Matthews has said, "Everything was new to them and each
new case presented problems for which there were no books to guide them, and no

Figure 2.   World War I casualties. (Reprinted from Bowlby, Sir A., and Wallace, C. The development of British surgery at the front. Chapter 6 in *British Medicine in the War, 1914–1917.* The British Medical Association: London, 1917, p. 33.)

articles to which to refer. With each case they created the methods with which to treat the next."[1]

The debt owed to Gillies by plastic surgery is enormous and cannot be lightly regarded. But in the specific area of flap design, Gillies, like Columbus before him, got it wrong and for the same reason. For Sir Harold's surgical flap concepts were grounded in the principle of geometric convenience as though the human skin were randomly supplied with blood. Such skin flaps were so precarious in their blood supply that they often required four, six, or more surgical stages and months of hospitalization for completion of even rather limited reconstructive tasks (Figure 4).

What Manchot would like to have told Dr. Gillies was that the human skin is not capriciously and randomly supplied with blood; rather, it is a precise mosaic of arterial territories supplied by highly predictable end-arteries, and that in the majority

1. Matthews, D. N. Gillies: Mastermind of modern plastic surgery. *Br. J. Plast. Surg.* 32:68–77, 1979.

Figure 3.   Sir Harold Gillies (1882–1960), the father of modern
plastic surgery (courtesy of Ralph Millard).

of cases, these vessels course deep to the muscle layer, sending perforating musculocutaneous branches arborizing to the skin above (Figure 5). If one wished to transfer blocks
of skin and soft tissue, it would thus seem prudent to respect this arterial distribution
rather than the mere whim of geometric convenience. And rather than divide these
critical perforating blood vessels surgically by separating the skin and fat from the underlying muscle, as was Gillies' wont when raising such a flap, it would seem essential to
preserve these connections and, in many cases, to transfer the muscle as well in a
single attached block. Gillies' long tubed skin pedicles and his other geometric flap
modifications were too randomly designed, Manchot might have argued, and could
predictably have resulted in a needlessly high rate of vascular insufficiency.

   For almost 60 years, Gillies' basic flap designs formed the framework for the
practice of four generations of plastic and reconstructive surgeons, always hovering at,
or too often just beyond, the circulatory limits of the small units of skin being transferred.
The realization that something of momentous importance had been embarrassingly over-

Figure 4.   Facial reconstruction at Sidcup in World War I. The many variations of the Gillies principles. (Reproduced with permission from Gillies H. D., and Millard D. R. *The Principles and Art of Plastic Surgery*. Little, Brown & Co., Boston, 1957, p. 36).

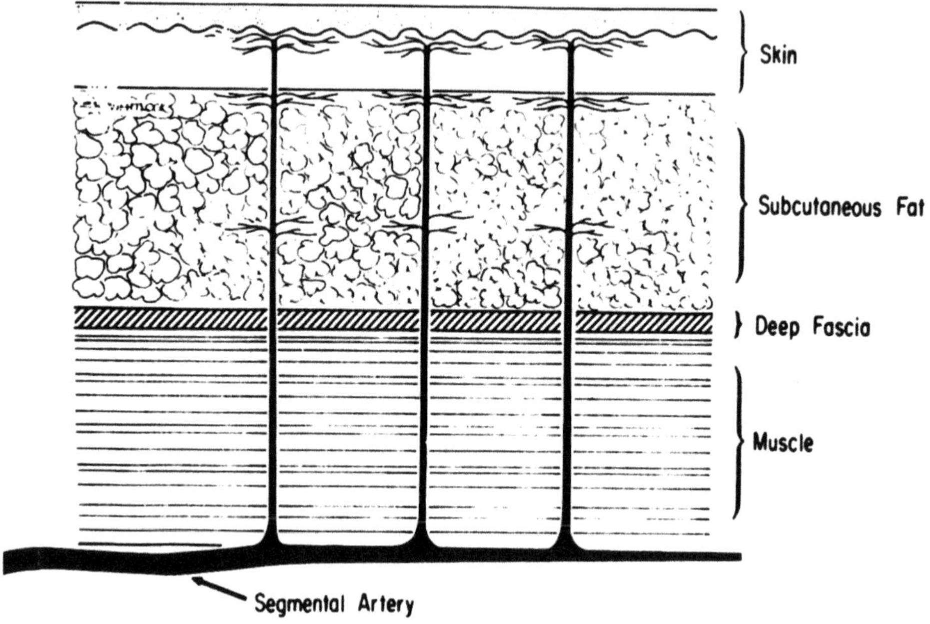

Figure 5. Musculocutaneous arteries originate from sizable submuscular segmental vessels. They perforate the muscle, fascia, and subcutaneous fat to supply the overlying skin. (Reproduced with permission from Morain W. D., and Salisbury R. E. Principles of plastic surgery. In Goldsmith H. S. (Ed.) *Goldsmith's Practice of Surgery*, Harper & Row, Philadelphia, 1979, p. 15.)

looked was only slowly admitted to the collective reconstructive surgical mind in the mid-1970s, resulting in a rather red-faced admission that it had been done all wrong for over half a century.

Once attention became properly refocused on the "arterialized" flap concept, the question spontaneously arose among the many investigators: What *is* the nature of the circulatory anatomy of the skin? Astonishingly, except in a few anatomic zones, no one seemed to know for sure, and standard comprehensive anatomy texts only skirted the subject. Postmortem rooms and medical school cadavers found new visitors as the answers were sought.

In the midst of this flurry of activity, Manchot's book was rediscovered, although very few copies of the German language work were available in American libraries. Photocopies from the National Library of Medicine in Washington began to circulate in some quarters, and one unpublished English translation from Oxford received some circulation. By plodding through the complex syntax of technical German or relying on the remarkably detailed illustrations, plastic surgery investigators could often document flap feasibility in advance and could plan new approaches on a rational basis. In the end, plastic and reconstructive surgery was born again, this time on a sound circulatory footing.

The geography of France adds slight irony to this footnote in surgical history. The millions of casualties sustained in that war that gave birth to modern plastic surgery occurred along an oscillating front, one end of which was never more than 100 miles from the Alsatian city on the Rhine where Manchot's anatomic studies had been performed. Columbus, an Italian sailor, had missed the critical message out of his own Greco-Roman heritage. Had the flood of casualties not been so overwhelming or had

Manchot published instead in English, Harold Gillies might well have encountered this important piece of background before proceeding haphazardly under the crush of wartime necessity. Had it been thus, it is doubtful such a creative surgeon would have missed the practical implications of the clear message gleaned from a nearby group of cadavers a quarter century before. And the entire history of plastic surgery would have been immeasurably altered.

—— II ——

Es kann doch unsereiner
Nur denken, wie er muss.
WILHELM BUSCH

Our kind can, after all,
Only think the way we must.

A book with the quality and scope of Manchot's does not arise *ex nihilo*. At first blush, it would seem an absurd notion that a young medical student in a structured curriculum could or would make the commitment to the tedious dissection and description of a large series of cadavers as well as the equally arduous task of preparing the text and illustrations for publication. One must understand not only the individual but also something of the background of the German academic tradition to see why this should have occurred.

The late nineteenth century was an ebullient period of German nationalistic fervor and optimism. Led politically by Bismarck and culturally and philosophically by Wagner, Brahms, Nietzsche, Marx, and Schopenhauer, the chauvinistic pride felt by Germans in all walks of life was well deserved and unquestioned. Industry, too, prospered with chemicals, machinery, and Daimler's new internal combustion engine. This ethnic awareness was a new phenomenon for Germany which had heretofore lacked the national consciousness long since enjoyed in France, England, and Spain.

At the beginning of the nineteenth century, the last Holy Roman Emperor, Francis II, had been overseeing a hodge-podge of central European feudal territories from his dubious throne. The Napoleonic Wars, summarily ending this medieval vestige, created the rather loose German Confederation that would prevail for the next half century. So humiliating was this defeat by Napoleon, however, that the consensual spirit of revulsion across the land became the spark from which the nationalistic spirit took flame. Fanned by the German romanticism of such writers as Ernst Moritz Arndt and Friedrich Jahn and later by the operas of Richard Wagner, the spirit soon became unquenchable. With a rising middle class in midcentury, Otto von Bismarck took charge of Prussian policy and skillfully melded German national sentiments into a strong central-ized government. In 1866, the year of Manchot's birth, the important victory in the Austro-Prussian War permitted the annexation of Schleswig-Holstein in the emergence of the strong North German Confederation. In 1871, the decisive Franco-Prussian War made Kaiser Wilhelm I the Emperor of Germany, now the strongest nation on the continent.[2] Alsace-Lorraine was ceded to Germany by the Treaty of Frankfurt, and

2. See Tembrock, R. *A History of Germany* (Trans. Paul J. Dine). Max Hueber, Munich, 1968, p. 223.

Figure 6.   Theodor Billroth (1829–1894) of Zurich and Vienna.

the Emperor chose to give his name to the three-century-old university in the Alsatian capital of Strassburg.

The university tradition shared the spirit. Theodor Billroth (Figure 6) wrote proudly in 1876:

> . . . since the fourth decade of this century, the German professor has been pretty generally required not only to know the results of the most recent researches, and to teach these to his students, but himself to be an investigator in the branch of science which he teaches. That combination had been appreciated in the earlier periods also, when happily it chanced to occur; but it is characteristic of the modern spirit of the German universities that they aim to be not only channels for conveying established knowledge, but also centers of research. They are to unite, to use a popular expression, "school and university" in one. Thus far, only the German universities have set themselves this high ideal as their objective; and in no other nation is it insisted upon as the most essential trait in the true character of a university . . . In every way, then, the German nation is making increasingly high demands on itself; it is inherent in the idealism that pervades the nation, that its striving for knowledge should be ceaseless, regardless of consequences, and intolerant of final dogmas . . . the task is difficult, but not too great for German energy.[3]

3. Billroth, T. *The Medical Sciences in the German Universities* (Trans. William H. Welch). The Macmillan Company, New York, 1924, pp. 27–28.

Our clinical instruction is purely German in form, and it is characteristically German constantly to find fault with it, to alter and improve it. As with most of our public institutions, we are conscious of its strong points, but we do not, like other nations, take any special joy in sunning ourselves in the glow of our achievements. We torture ourselves in the endless striving for constant improvement, in searching out our weaknesses and formulating plans for betterment. Thus, our cultural endeavors exhibit, like our politics, an eternal restlessness, an apparent incertitude, and bear, despite all our idealism, a pessimistic stamp which is unintelligible to the French and the Italians and intolerable to the English.[4]

Billroth was not using hyperbole. Whereas by 1830 nearly all Germans could read and write, only 60 percent of Britons and 40 percent of Frenchmen were able to do so as much as 35 years later.[5] The German educational tradition, long surpassing that of any rival, achieved its expected apogee through the richly endowed state university system. The very pride that built these institutions would also maintain their excellence through the rivalry for great scholars among the universities of neighboring states. In addition, three institutional pillars of academic freedom were established and stand yet as the foundation for independent scholasticism: (1) academic self government through professors and deans; (2) *Lehrfreiheit* (freedom of teaching) whereby a professor was enabled to teach whatever he wished and stand unquestioned in his convictions; and (3) *Lernfreiheit* (freedom of study), which freed students to attend lectures of their choice and to migrate from one university to another, unrestricted by rigid curricula, and responsible only to their examinations at the end.[6]

The academic tradition in German medicine had also attained formidable stature over these decades, the basic sciences in particular remaining unsurpassed in the world. Rokitansky, Virchow, and von Recklinghausen (Figure 7) founded modern pathology. Koch systematized the role of bacteria in disease. Ludwig and Volkmann led the new schools of physiologic investigation. The science of ophthalmology was commenced by Beer and Horner. Internal medicine was advanced through the efforts of Schönlein, Romberg, Krukenberg, Henoch, and Kussmall. Roentgen would soon introduce the X-ray. And the surgical honor roll was studded with the names of Billroth, von Langenbeck, Richter, Dieffenbach, Thiersch, Esmarch, and Trendelenberg. Of equivalent importance with the excellence of these teachers, however, was the rigorous system of postdoctoral residency training that was to be transplanted first to the Johns Hopkins Hospital by William Halsted and was later to become the standard for the remainder of American medical education.[7]

The study of anatomy was especially well adapted to the rigorous analytic nature of the German dialectic. A listing of the prominent German anatomists of the previous century reads like an eponymic index of *Gray's Anatomy:* Meckel, Henle, Meissner, Waldeyer, Leydig, Lieberkühn, Treitz, Langer, Heidenhain, Wolff, Scarpa, Rathke, Müller, Schwann, and Küpfer. Yet there was substantial disagreement and rivalry among the various schools and traditions across the country, and every student of anatomy could trace his academic pedigree back through several professorial generations.

4. *Ibid.*, p. 68.

5. Craig, G. A. *Germany 1866–1945.* Oxford University Press, New York, 1978, p. 187.

6. *Ibid.*, p. 199.

7. See Rutkow, I. M. William Stewart Halsted and the Germanic influence on education and training programs in surgery. *Surg. Gynecol. Obstet. 147*:602–606, 1978.

Figure 7.   Friedrich D. von Recklinghausen (1833–1910). This bust stands at the Institut d'Anatomie, Université Louis Pasteur, Strasbourg, France.

In the Germanization of the Kaiser Wilhelm University at Strassburg after 1871, the prominent neuroanatomist Gustav Schwalbe of Jena was given an important professorship in the Anatomy Institute (Figure 8). Schwalbe was a product of the South German school, tracing his pedagogical lineage through Max Schultze, Schultze (the elder), Ignatz Dollinger, Lorenz Oken, Carl Kielmeyer, von Siebold, Barth, Scarpa, Wolff, Meckel, and the great Albrecht Haller.[8] As we shall see, Professor Schwalbe was to play the key catalytic role in the production of Manchot's book.

8. Billroth, T. *Op. cit.*, pp. 216–219.

Figure 8. Gustav Schwalbe (1844–1916). This metal relief hangs
in the Institut d'Anatomie, Université Louis Pasteur, Strasbourg,
France.

III

In freier Luft, in frischem Grün,
Da, wo die bunten Blümlein blühn,
In Wiesen, Wäldern, auf der Heide,
Entfernt von jedem Wohngebäude,
Auf rein botanischem Gebiet,
Weilt jeder gern, der voll Gemüt.
Hier legt sich Bählamm auf den Rücken
Und fühlt es tief und mit Entzücken,
Nachdem er Bein und Blick erhoben:
Gross ist die Welt, besonders oben!

W. BUSCH

In open air, in Nature's freshness,
Where colorful little flowers bloom,
In meadows, forests, heathland
Far from any residential building
In a purely botanic area,
That's where anybody full of feeling likes to be.
This is where Baalamb lies down on his back
And feels deeply and with delight,
After having lifted his legs and eyes:
The world is great, especially above!

Figure 9.   Manchot's birthplace near Zurich. His father was pastor of this church at Wipkingen (courtesy of Gertrud Manchot).

Carl Manchot was born in Wipkingen, Switzerland, near Zurich, on April 30, 1866 (Figure 9). His forebears were of Huguenot extraction, having fled French religious persecution several generations before. As the firstborn, he was named after both his father and his maternal grandfather, while his twin sister, Caroline, took her mother's name.

Education and religion were the distinctive features of the family heritage. His father, Karl Hermann Manchot, possessed a rare doctorate in theology from studies at Giessen and at the University in Zurich and was a Reformed Protestant minister in the local parish near his own former seminary at the time of Carl's birth (Figure 10). The Reverend Manchot's father, Johann Daniel Manchot, in turn, had been a theologian of high rank in Offenbach, Germany, near Frankfurt (Figure 11). Carl's maternal grandfather, Professor Carl Credner, also a prominent Protestant theologian, had married the daughter of Heinrich Luden, a famous professor of history at the university in Jena. Professor Luden, Carl Manchot's great-grandfather, was a contemporary of both Schiller and Goethe and was once cited by the latter in a famous published conversation with Eckermann.

The immediate family was traditional with the single exception that Carl's father had chosen a woman seven years his senior for a wife, a highly unusual event in the patriarchal society of the day. Caroline Credner Manchot, 33 years old at the time of Carl's birth, had a cheerful and informal disposition, striking a balance with the more austere Calvinistic countenance of her husband (Figure 12).

Figure 10. Karl Hermann Manchot, Carl's father (courtesy of Gertrud Manchot).

Figure 11. Johann Daniel Manchot, Carl's paternal grandfather (courtesy of Gertrud Manchot).

Figure 12. Caroline Credner Manchot, Carl's mother (courtesy of Gertrud Manchot).

Figure 13.   Carl with his younger sister Anne (*left*) and his twin sister Caroline (courtesy of Gertrud Manchot).

The family traditions in advanced education were inculcated from the beginning with all of the children, including the girls. Carl's brother Wilhelm ultimately became a Doctor of Pharmacy in Munich while his twin sister received a broad education in the humanities. His sister Anne died of tuberculosis in her teens, and his brother Walter died in infancy.

Carl and Caroline Manchot were brought onto German soil at six months of age when their father was called to be associate pastor of the St. Remberti Church in Bremen to serve alongside the Reverend Johann Kradolfer, a fellow Zurich seminarian. St. Remberti was one of the city's oldest and most well-established churches, dating from the height of Bremen's commercial power as a Hanseatic city on the North Sea in the fourteenth century. Originally with the Roman Church, St. Remberti had become Protestant in 1524, just seven years after Martin Luther had tacked his 95 theses to the door of a church 100 miles to the east. The Reverend Karl Manchot would be the twenty-fourth Protestant pastor in consecutive service.

For reasons already noted, 1866 was a propitious year to take up residence in northern Germany. Bremen shared the prosperity of the North German Confederation as the second largest seaport of the region. The church coffers swelled, as did the Manchot family (Figure 13), and the Reverend Manchot found that the increase would permit an expansion in the Lord's work. Cramped in a deteriorating 130-year-old boxlike building, the parishioners hired an architect and built a splendidly large and multispired Gothic structure that became both a scenic and cultural center of the city (Figure 14). An energetic and socially conscious leader of the flock, the Reverend Manchot instituted a parish newspaper and also founded and supervised a community home for the elderly and a nursery school on adjacent land.[9]

9. The Church itself was totally destroyed in a 1942 bombing raid, but the nursing home and the parsonage where the Manchot family lived are still standing and in use.

Figure 14 *Left:* St. Remberti church in Bremen. The Reverend Karl Manchot supervised the construction of this cathedral in 1871 (photo from the rectory of the present St. Remberti church). *Below:* Parsonage of St. Remberti where Carl lived as a child.

Figure 15.   St. Gertrud Lutheran church in Hamburg.

Although young Carl must have been deeply affected by his father's humanitarian concerns, his own aptitudes and interests drew him not toward theology but toward natural science. From an early age he developed a fascination for the outdoors and especially for wildflowers. He soon learned them all by name, first in German and then in Latin, and the tiny Veronica would remain forever his favorite. It was at this time that he developed a passion for his lifelong favorite leisure activity—the botanical expedition. Indeed, his twin sister was repeatedly dismayed and embarrassed when he chose to spend his Sunday mornings in the woods rather than in the family pew listening to their father.

In 1883, when Carl was 17 years old, the family moved east to Hamburg which was to be, with notable short exception, the doctor's home for the remainder of his life. His father, then 44 years of age, had accepted a new position as pastor of the large St. Gertrud Lutheran Church (Figure 15). This institution was even better suited than St. Remberti to Carl's socially activist father, since it sponsored a home for the elderly, an orphanage, and several other allied humanitarian activities over which the Reverend Manchot would exercise coordinating supervision as pastor.

We may guess that young Carl's parents still carried a wish at this time that he follow the long family tradition in the study of theology, for he was enrolled in the prestigious Classical School of Johanneum. Although his academic performance was sterling indeed, it is perhaps of greater significance that he enrolled himself at the Oberrealschule Arngardstrasse of natural science for special after-hours courses of his own choosing in addition to his classical studies. After two such years, the logical compromise of a medical career must have been an acceptable one for all concerned.

Thus was the long theological tradition broken in the Manchot family, or at least transformed into the humanitarian alternative of a career in medicine. If the Reverend Karl Manchot would not see any of his progeny taking up the cloth of his fathers, he might at least have taken solace in the knowledge that 12 of his grandchildren and great-grandchildren or their husbands would continue to serve mankind as doctors and dentists; and it was Dr. Carl Manchot who was to be their role model.

——— IV ———

Mensch mit traurigem Gesichte,
Sprich nicht nur von Leid und Streit.
Selbst in Brehms Naturgeschichte
Findet sich Barmherzigkeit.

W. BUSCH

Oh man, with the sad look on your face,
Don't talk just about suffering and strife.
Even in Brehm's Natural History,
It is possible to find charity.

On October 28, 1885, at the age of 19, Manchot matriculated at the Kaiser Wilhelm University in Strassburg, signing the admission roster in his bold longhand script and designating his course of study as Medicine (Figures 16 and 17). (Harold

Figure 16.   Kaiser Wilhelm University, now Université Louis Pasteur, Strasbourg, France.

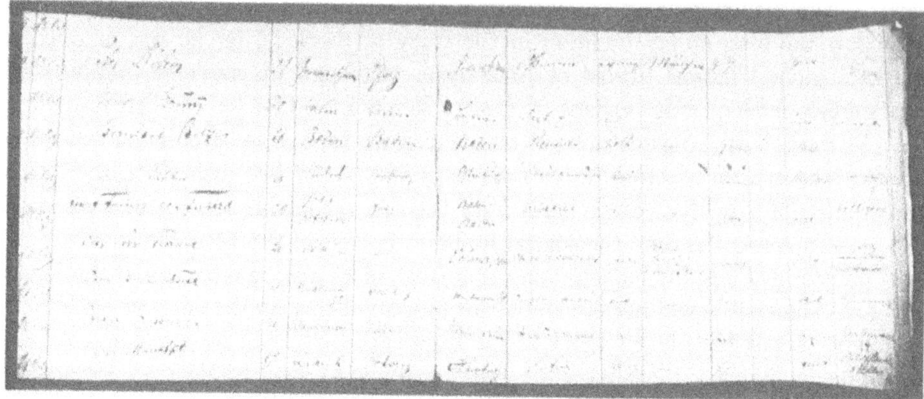

Figure 17. Manchot's signature on matriculation at Kaiser Wilhelm University, designating his intention to study medicine (from the archives of the Institut d'Anatomie).

Figure 18. Entrance to the laboratory where Manchot's dissections were performed.

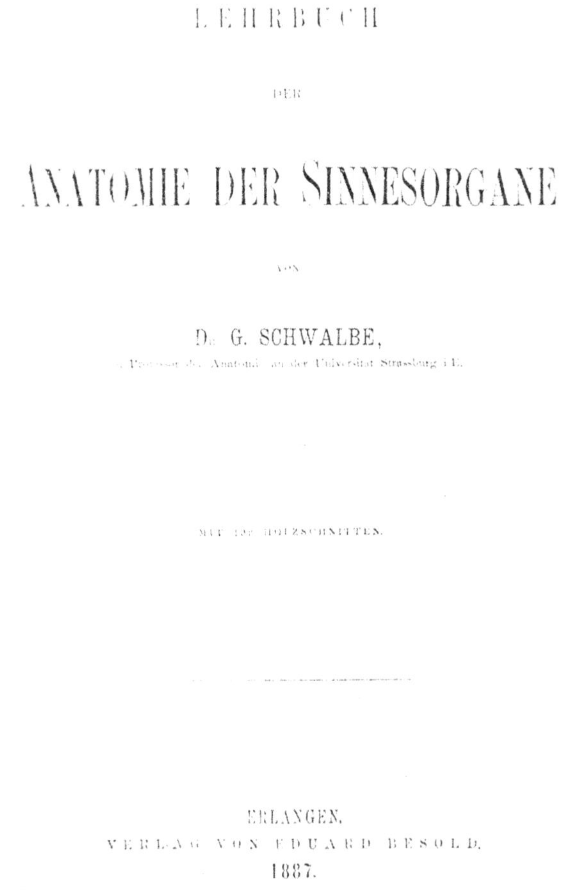

LEHRBUCH

DER

ANATOMIE DER SINNESORGANE

VON

Dr. G. SCHWALBE,

Professor der Anatomie an der Universität Strassburg i. E.

MIT 132 HOLZSCHNITTEN.

ERLANGEN.
VERLAG VON EDUARD BESOLD.
1887.

Figure 19. Title page of Schwalbe's principal work, from which inspiration for the dissections arose.

Gillies was then 3 years old and 14,000 miles away.) There is no suggestion in the available record why young Carl should have chosen this institution with French flavor so far from the familiar North Sea ports where he had been raised. No family members and friends were known to have been in Alsace, and the school itself was not noted for outstanding faculty or other particular distinction, at least not in Billroth's contemporary dissertation on German medical education.[10] Manchot's daughter Gertrud has speculated that the government under Bismarck might have offered financial inducements for German students to enter these newly annexed and Germanized universities. If so, there is reason to believe that this would have carried substantial weight in the Manchot parsonage in Hamburg.

It was during his freshman year, while leading his class scholastically, that Manchot's nascent professional career became intertwined at the Anatomy Institute with that of Professor Dr. Gustav Schwalbe, whose academic roots we have already traced (Figure 18). By chance, the prolific neuroanatomist was just going to press with a comprehensive textbook on the anatomy of the sensory organs[11] and was fully primed on the location of the frontiers of knowledge in his own field (Figure 19). Schwalbe was thus

10. Billroth, T. *Op. cit.*, pp. 169–182.
11. Schwalbe, G. *Lehrbuch der Anatomie der Sinnesorgane.* Besold, Erlangen, 1887.

in a favorable position to outline profitable research areas at this time and was especially eager to pursue some of them. But because, like so many of his colleagues, his teaching and writing duties made such overwhelming demands on his time that his ability to conduct his own bench investigations was extremely limited, he was always alert for opportunities to enlist junior collaborators.

Schwalbe had observed that a few of his university colleagues had managed to advance their own research by proposing its completion for the annual student prize. The 300 mark stipend could usually attract interest among the student body. Indeed, it had long been the custom at the Kaiser Wilhelm University—not only in the medical school, but also in such schools as law, natural science, mathematics, and philosophy— to award an annual competitive student prize for a paper on a specifically assigned subject. But what would be a suitable project for a preclinical medical student, one upon which the faculty committee would seize as an especially meritorious suggestion for a student project?

Schwalbe remembered writing in his book of the dearth of knowledge concerning the neural interconnections of the specialized cutaneous sensory endorgans, the Meissner and Pacinian corpuscles. On page 6 of his book, in referring to cutaneous nerve fibers, he had written:

> Unfortunately, more exact quantifications of the abundance of nerve fibers in individual cutaneous regions are not available, determination of which would be of the greatest importance with respect to questions of fundamental physiology . . . . Another fact worthy of research for the same reason is the encroachment phenomenon of the "plexus-territory" of one cutaneous sensory nerve onto the territory of another. How substantial this encroachment could be has not yet been ascertained.

But tiny nerves in the subcutaneous fat could be difficult to trace, even for a trained microscopist; that is, unless they happened to course with the more easily identifiable cutaneous arteries which were at least injectable. Would a study of these cutaneous vessels open the book on the entirely unknown area of terminal cutaneous nerve distribution? And if it did, could not these arteries, and therefore the nerves, be tied anatomically to the primitive segmental arrangement of the spinal cord and therefore of the embryologic somites? Could the complexity of the nerve and arterial distribution to the head, trunk, and extremities be joined into some unifying *Gestalt* concept of somitic origin by such an investigation? Clearly, no limited regional study was in order; if such a study were to be undertaken, the integumentary covering of the entire body had to be examined.

Schwalbe laid out the prospective assignment with its arterial emphasis to make the project an attractive and feasible one for a student project:

> A summary of the cutaneous arteries of the human body should be given according to origin and distribution with comparison to the distributional territories of the cutaneous nerves and taking into consideration the metamerism of the body.[12]

His grand aspirations of a unified embryologic theory prevailed over other entries, and the faculty committee accepted his project for the 1885–1886 academic year.

Schwalbe was not really interested in the skin arteries at all, and herein lies another irony from the standpoint of plastic surgery history. As a neuroanatomist, he

---

12. Institut Anatomie. *Archival Record,* 1886, Louis Pasteur Universite, p. 12.

viewed the arteries as merely the template on which he hoped to read and decode the secret of the cutaneous nerve distribution. The important conclusions, he hoped, would be a unifying thread between the fields of neuroembryology and neuroanatomy. (One wonders how irrelevant Harold Gillies would have found such a line of questioning to be on receiving his first detachment of maimed infantrymen at Aldershot in 1916.)

But the assignment was to prove more challenging than originally believed by Schwalbe and the faculty committee. In fact, when it became apparent that no acceptable entry would be received during the entire year, the proud professor must have seen his grand hopes dwindle for his unified neuroembryologic theory. How could he return to the faculty committee and explain his personal failure to inspire the enthusiastic execution of such a pivotal project?

Carl Manchot was to provide Schwalbe a way out of his predicament. Near the end of the school year, the young student approached his favorite anatomy professor concerning the competition. He suspected that it was Schwalbe himself who had proposed the specific assignment and wondered if there were sufficient time yet to complete it. Knowing of Manchot's outstanding scholarship and having seen the student's excellent anatomical sketches of his routine dissections, the professor sensed that his research dream might yet be saved, but only if he could buy more time. The project would likely require months of work for someone as meticulous as Manchot.

Schwalbe took two actions. He successfully petitioned the faculty committee to extend the assignment another year, justifying his request by assuring the faculty committee that he was taking an appropriate measure to assure its satisfactory completion. This second measure took the form of a letter to the Dean of the University in August of 1886 on behalf of his young protégé. This was a request for a funded assistantship at the Anatomy Institute for the 6-month period from October 1886 to April 1887. During this interval, Manchot would largely withdraw from his regular classes and laboratories and would instead devote all his energies toward completing his research and entering the competition. For a stipend of 25 marks per month, he could hardly do less.

As Manchot eagerly commenced his assignment, however, it became apparent that those prized cutaneous nerves of Schwalbe's grand design simply could not be identified in his cadaver preparations. Therefore, the only data that could reasonably be studied would come not from inferences about nerves he could not readily find. He must instead focus entirely on the exact course and distribution of the arteries themselves. They at least were injectable. If there were speculations to be made about accompanying nerves or presumptive somitic derivations, they must follow, not parallel, the arterial dissections. As he precisely identified his priorities in the foreword:

> The assignment was to fill existing gaps in the knowledge of the arterial supply of the skin. These assignments further requested a comparison of the cutaneous arterial and nerve supplies and an investigation of the interrelationships which might exist between the arterial distribution and the somitic arrangement of the body, thus leading to an understanding of the basic organization of the skin arteries.

The methods of the investigation are only sketchily known. Nowhere in the book does he catalog his study materials nor detail his special dissection techniques, and no records concerning these points are to be found in the Archives of the Anatomy Institute of what is now the Louis Pasteur University in Strasbourg, France. How many cadavers were studied is unknown, but the detail in the text would suggest that the number was sizable. On several occasions in the text, he does make direct reference to injection of the arteries. While it is unclear what materials were used, India ink or some form of latex would be strong contenders. Nevertheless, the aniline dye industry was bringing great prosperity to several German chemical companies at this time, and we may speculate that these many brilliant options in color could have provided the sharp delineations in vascular territory and the numerous anastomotic relationships he frequently describes in his text.[13] (Indeed, as a further historical irony, it is these very pigments that plastic surgeons now use to sketch the outlines of their arterialized flaps on the skin.) Injection or no, it is nevertheless well known to all who have attempted to trace cutaneous end-arteries in even latex-injected fixed specimens that it can be a frustrating experience. To have included the entire human body in the scope of the investigation seems such a formidable undertaking that one wonders at the wisdom of such an awesome assignment. But a regional delimitation could scarcely have answered the cosmic questions motivating Dr. Schwalbe and, in turn, his industrious student.

Manchot was clearly equal to the task. His strong taxonomic background in botany had given him an appreciation for categorization, classification, and order. As already noted, his meticulous sketches from his basic gross anatomy class[14] had caught the eye of his professors and classmates alike. For the drawings that were to accompany the manuscript in its final submission for the competition, Manchot traced the muscle diagrams from two popular general anatomy texts of the day, viz. those of Gegenbaur[15] and Sappey,[16] and overlaid the arterial patterns in red (Figure 20).

A simple description of the courses and relations of the arteries was insufficient for the larger purposes of the original assignment, however. Some unifying synthetic leap was required. Although it was scarcely the neuroembryological one for which his mentor would have wished, the young naturalist must have had a flash of insight during his reflections that produced precisely the idea that never occurred to Gillies despite a long and brilliantly creative career working on much the same problem from another direction. Manchot's illuminating insight, like that of his much more famous fellow young anatomist of $3\frac{1}{2}$ centuries before, lay not just in descriptive detail but in the conclusions those details had revealed to their observer. What all those individual arteries finally conveyed to Manchot was the revolutionary notion of cutaneous arterial "territories"—that the skin of the human body is not supplied by a haphazard rete of random small vessels but is instead organized into a predictable mosaic of regions, each deriving its blood supply from a specific named arterial source. Manchot identified these territories, defined their limits, and mapped them on front and back diagrams of the human body.

These summation drawings, along with the labeled arterial sketches and the book-length manuscript, were submitted as Manchot's entry in the student competition. He was unanimously awarded the grand prize along with its accompanying 300 mark stipend. And as special recognition for the exceptionally meritorious original contribution the work represented, the unawarded 300 mark stipend of the previous year was added as a bonus.

13. Were this in fact the case, those important aniline dye flush injection studies of Kaplan at Stanford and others in the late 1960's would actually have been modern repetitions of some of Manchot's original work.

14. These sketches, now in the possession of a physician nephew in Rostock, East Germany, show a precision in detail which is reflected in the later book illustrations.

15. Gegenbaur, K. *Lehrbuch der Anatomie des Menschen*. Engelmann, Leipsig, 1883.

16. Sappey, M. P.H.C. *Traite d'Anatomie Descriptive*. Delahaye, Paris, 1876.

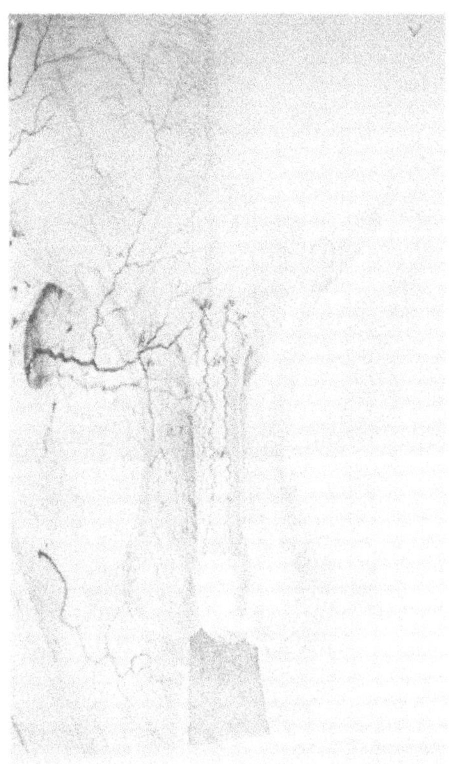

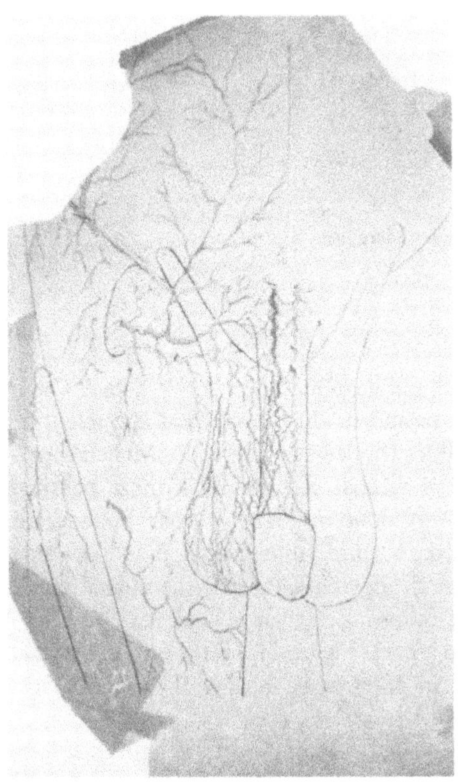

Figure 20 *Above left:* Pencil sketch on brown paper, probably representing Manchot's first sketch of the superficial inferior epigastric artery (*center*) and the superficial circumflex iliac artery (*extreme left*) (courtesy of Gertrud Manchot). *Above right:* Intermediate diagram on parchment of the same vascular territories (courtesy of Gertrud Manchot). *Left:* Final finished diagram on white paper. It was in this form that the illustrations, all personally drawn by Manchot, were submitted to the faculty committee in consideration for the competition prize (courtesy of Gertrud Manchot).

Figure 21. Apartment building in Strasbourg where Manchot
lived during his medical studies.

Immediately, Professor Schwalbe urged publication. The F.C.W. Vogel Company of Leipzig, then one of Germany's most prominent in the medical and scientific field, accepted the monograph, provided the illustrations were upgraded. Manchot persuaded his artist friend Bamberger to redo the arterial sketches in a more polished form. The two important summary drawings were done in color by a Mr. Schneegans whose identity cannot be further characterized. A third summary diagram depicting the lateral projection was considered but eliminated, apparently for space considerations.[17] It is likely that Manchot spent much of the following regular school year of 1887–1888 readying the manuscript for submission to Vogel (Figure 21). Final publication did not come until his return from study abroad in 1889 when he was 23 years of age.

17. Such a preliminary lateral sketch was found among Manchot's papers in the possession of his physician daughter in Bamberg, Bavaria.

There are several imaginable motivations for Manchot's decision to undertake his important study in the first place, but one can only speculate about these. Certainly the heart of the naturalist must have been stimulated by the opportunity to explore an uninvestigated area of biological science, and it must be assumed that some genuine personal curiosity existed concerning the nature of his subject. Furthermore, the personal affection between professor and protégé seems also to have been a potent stimulus.

Nevertheless, when one considers the financial constraints of the Manchot family, it is altogether possible that the pecuniary interest in the competition's prize was a certain spur toward achievement. He cannot have been alone in this for, as Billroth points out, "the German medical profession is recruited more largely from among the sons of ministers, of petty officials, of schoolteachers—in short, from social classes that do, indeed, lead an economically circumscribed existence, but whose members in general nevertheless belong to the educated classes."[18] Billroth put the total cost of a medical education through the first 6 years of practice at 24,000 marks, ". . . based on the assumption that the student will not early in his career commit the folly of marrying a poor girl."[19] Scholarships and prizes must have been sought to lighten this burden.

On another level, although psychohistory is a treacherous discipline at best, one could even speculate that the desire to placate the perceived expectations of his austere, achievement-oriented father played an offstage role of some significance. Recognition for a major academic achievement would surely help alleviate such inward pressures, especially if spiked with any guilt over having made a second-best career choice.

Alternatively, since there is some evidence that Manchot was not fully satisfied with his Kaiser Wilhelm University experience, it is possible that undertaking the work was a planned diversion from the daily routine. A second such diversion had already been sought the month before commencing his assistantship at the Anatomy Institute when Manchot traveled to his birthplace outside Zurich with introductory letters in hand from his father. Taking advantage of *Lernfreiheit* (the "freedom of study" tenet of the German university tradition described earlier) during this short visit abroad in 1866, he made preliminary arrangements to complete his preclinical studies at the University of Zurich.

Following completion of his research in April 1887, Manchot returned to his preclinical studies for another year before departing for Switzerland. He spent only 12 months at the university in Zurich but fell utterly in love with the mountains during his stay. Nor did his wildflower field trips cease, for with his newly found and lifelong friend, Dr. Ris, he made many lengthy expeditions into the mountain trails above the city. After one such hike, they were noticeably conspicuous as they tried to slip into the lecture hall well after the appointed lecture had commenced. The professor, in a fit of Teutonic pique, stood them in front of the class and introduced Ris as "sunburn, second degree" and Manchot as "stupidity, third degree." Manchot rounded out his activities by singing in the University chorus at Zurich and returned to Strassburg in April 1889 to begin his clinical studies. The skin arterial research manuscript entered publication in book form the same year.

During the course of Manchot's final 2 years of medical study, he produced one more significant piece of investigation, a doctoral dissertation on the origin of true aneurysms. This exhaustive treatise was regarded as so important by his famous pathology

18. Billroth, T. *Op. cit.,* p. 108.
19. *Ibid.,* p. 151.

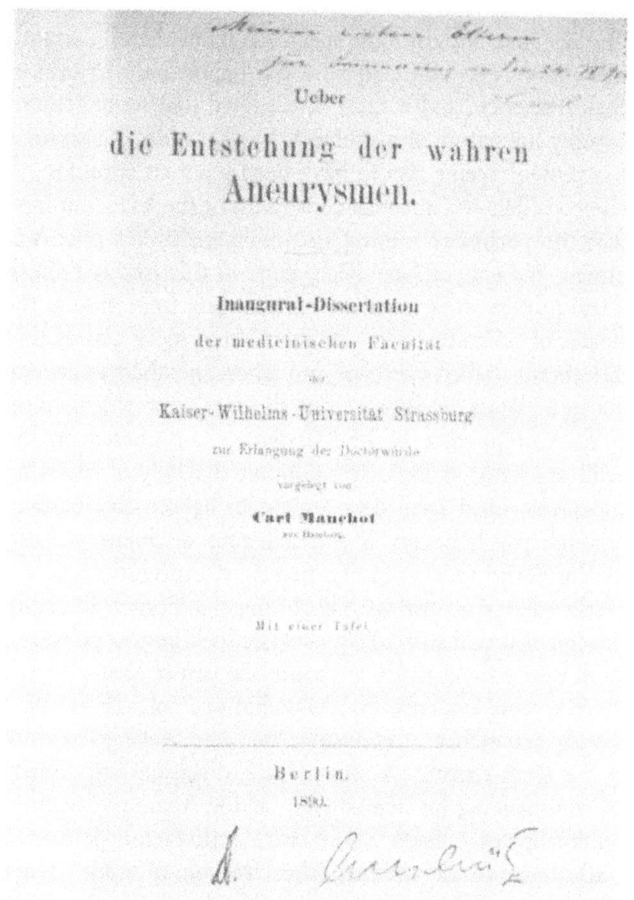

Figure 22. Title page for Manchot's published dissertation. Note autograph at top (courtesy of Gertrud Manchot).

professor, von Recklinghausen, that the professor sponsored its publication in the prestigious *Virchow's Archives* (Figure 22).[20] Manchot took his degree as Doctor of Medicine in time for his parents' silver wedding anniversary and passed the Alsatian state board examination the following year.

—— V ——

Nun klopf ich ganz bescheiden
Bei kleineren Leuten an.
Ein Stückel Brot, ein Groschen
Ernähren auch ihren Mann.
W. BUSCH

Now I very modestly knock
On simpler people's doors.
A slice of bread, a penny
Are enough to subsist on.

20. Manchot, C. Ueber die Enstehung der wahren Aneurysmen. Virchows *Arch. Pathol. Anat.* Berlin, 1890, p. cxxi.

Dr. Manchot was now at an important junction in his life. Should he follow an academic career tradition, or should he follow his more humanitarian calling into private clinical practice? His dilemma was compounded by the fact that he was unquestionably well suited for both options.

A career as a professor must have seemed very attractive at this time. His idolatry of Schwalbe and von Recklinghausen and the very high esteem in which university professors were held in German society would have been strong motivating factors in this direction. His father, too, would undoubtedly have taken pride in such a career. There was little doubt that Manchot was of professorial caliber if one judges from the remarkable quality of his book and dissertation. For financial support, a system of funded university assistantships and lectureships had been established in 1875 by the Prussian Minister of Education to enable such promising young academicians to survive their early academic years until they could receive regular appointments.[21] Further, he had no financial family obligations of his own at this time and could have made the continuing necessary personal sacrifices to achieve his goal.

But how great were those sacrifices to be? Financial security aside, open university positions in basic science were rarer than the most exotic of Manchot's wildflowers. Again, Billroth laments:

> . . . what is to become of all the candidates for professorships in anatomy, physiology, and pathological anatomy that are produced every three or four years, and for whom there are in German Austria precisely three places? What is to be done with them? . . . In the German Empire, there has been in these fields such an overproduction for several years that even emigration to Italy, Holland, England, and Russia has not been able to relieve the pressure of crushing competition . . . .[22]

> The best motto for the private lecturer to keep in mind is the sentence by Plater, quoted by Jacob Grimm: "Then I took up my seat in a corner not far from the schoolmaster's chair and I said in my heart: In this corner you are going to learn, or die."[23]

And thus it must have appeared to Manchot in 1891. His desire to return home to Hamburg and the North Sea held sway over such a dubious future. Perhaps a suitable compromise could have been reached in a clinical specialty, for learned private practitioners, then as now, offered time in university clinics to support the educational structure. But as Hamburg had no medical school, this option was not possible. It is difficult to imagine, though, that all academic aspirations had left Manchot's mind when he accepted his first postdoctoral hospital position.

On the other hand, it must be remembered that Manchot's tutelage had been steeped in a tradition of humanitarian concerns, and he had seen at first hand the hardship of the poor and elderly and especially the disadvantaged children. His reexposure to these ubiquitous problems during his clinical studies would surely have unleashed the compassionate motivations that had compelled him toward a career in medicine in the first place. Whatever remorse there was in leaving the cloister of academia was to be swallowed up in Manchot's conscientious commitment to his clinical responsibilities.

21. Billroth, T. *Op. cit.*, p. 210.
22. *Ibid.*, pp. 214–215.
23. *Ibid.*, p. 211.

Figure 23.   Eppendorf Hospital in Hamburg, site of Manchot's first postdoctoral appointment.

Dr. Manchot took a position as medical resident at the Eppendorf Hospital (now the University Hospital) immediately on his return to Hamburg (Figure 23). He had not been there a year when he faced a major clinical challenge.

In the summer of 1892 Hamburg was struck by a cholera epidemic. Although it had been nearly four decades since John Snow had first put an end to such an epidemic by removing the handle from the Broad Street pump in Soho and eight years since Manchot's countryman Robert Koch had isolated the *Vibrio comma* bacterium in Berlin, the plague mentality among the public at large was still extant. The populace fled to the countryside and the city's hospitals overflowed with the mortally dehydrated. Manchot made certain that all of his family members were safely evacuated as he offered his services at the center of the epidemic (Figure 24). It was the blend of scientist and humanitarian that permitted him to write a touching letter to his family at the beginning of their exile, instructing them exactly on the things to take with them and those to leave behind, and reassuring them that his scientific judgment would protect him as he carried out his responsibilities. All of the family survived and were reunited in the fall.

Three years later Manchot accepted a new position as assistant physician at St. George Hospital in the Uhlenhorst District and apprenticed himself to Dr. Engel-Reimers, a prominent but eccentric dermatologist. What Schwalbe had been to Manchot as academician, Engel-Reimers now became as clinician. The dermatologist was a rough study, proud of describing his rude background as the illegitimate son of a washerwoman. Engel-Reimers was an outspoken and good-natured bundle of maxims, jokes, and anecdotes as he often made his more socially prominent colleagues appear stiff and pompous. On more than one occasion, for example, he had been known to leave a patient's

Figure 24.   Manchot and his colleagues during the Hamburg cholera epidemic in 1892. He is
the iconoclast leaning on the post in the front row (courtesy of Gertrud Manchot).

sickroom with long face and to drone to the mourning-clad family in the most somber
of tones, "You must be prepared to face the worst . . .," adding after a hushed pause,
"he will survive." Manchot quickly perceived the value of the quick wit at the bedside
and tried to emulate his mentor the best that the son of a Calvinist clergyman could.

Shortly after his arrival in the Uhlenhorst, he began to notice with more than
passing interest an attractive young woman living in a house across the street (Figure
25). As he idled down the front walk each morning, he admired the way she made
sandwiches at the kitchen window for her younger brothers and sisters. Since Emmi
Cropp's father Caesar had died when she was 15 years old, this was but one of her
many family responsibilities as the eldest child. The Cropp family, Manchot came to
learn, had been prosperous Burgermeisters in Hamburg for several generations and were
well placed in society. Nothing typified the commercial success and spirit of the Hanseatic
leagues more than the wholesalers and shippers such as Emmi's father. As the young
doctor's interest in the oldest Cropp daughter grew more and more ardent, the gnawing
question of the difference in social background became increasingly important. A proposal
of marriage was insufficient; the arrangement could be consummated only if the apparent
discrepancy in social station could be resolved. A letter to the Cropp family from the
Reverend Karl Manchot soon arrived, detailing the excellent background of distinguished
theologians and philosophers in the suitor's family. They were married in 1898 (Figure
26).

Dr. Manchot severed the remaining vestiges of his apprenticeship the year
after his marriage and settled into a blossoming private practice of family medicine.
Far, far from the exploratory leisure of the dissecting room, he made his daily housecalls
by horse-drawn buggy with black bag at his side. He

Figure 25. Emmi Cropp, Carl's bride-to-be (courtesy of Gertrud Manchot).

Figure 26. Dr. and Mrs. Carl Manchot about 1900 (courtesy of Gertrud Manchot).

Figure 27. Manchot with raison d'être (courtesy of Gertrud Manchot).

. . . used to observe long and thoroughly at the sickbed and then give his instruction in few words. Through his calmness—perhaps sometimes too comfortable a calmness—he had an overridingly positive and trust-inspiring influence on his patients as well as their concerned parents. Professor Manchot won a faithful and thankful clientele not only as a good family doctor but also as a wise family advisor who always had a fine sense of humor.[24]

But this was not quite enough for a man who relished the idea of a continuing personal project.

Dr. Manchot loved children, and the children instinctively recognized his sympathetic good nature (Figure 27). As the large families of the day assured there would be so many youngsters about, the lion's share of his family practice gravitated happily into pediatric care. It was this sense of personal enjoyment that, coupled with the respected memories of his father's good works, compelled him to seek the medical directorship at Hamburg's orphanage in 1905, a task that would become the all-consuming nucleus of his career over the next 25 years.[25]

24. Oehlecker, Professor. Zum Gedächtnis (A Eulogy to Carl Manchot) (Trans. D. Morain). *Norddeutches Aerzteblatt,* October 16, 1932.
25. Dr. Gillies was at the same moment an intern at St. Bartholomew's Hospital in London.

Figure 28.   Orphanage designed and built by Manchot in Hamburg. It is still in use today as a day-care center.

There were to be many facets of this challenging position. Not only did Manchot become responsible for the medical supervision of the orphans in the home; he also oversaw the entire foster children's program, supervised the young girls under public relief, and soon inherited the thousands of war orphans ballooning the rolls. His initial half-day commitment overflowed into many, many evenings to keep up with his paperwork. Every morning he found his desk and chair piled precariously with thick, heavy files. His daughter recalls that in the busy early 1920s, if he missed a day's work, he would find the following day two more chairs brought into the office and stacked 2 feet high each with more of the folders.

Manchot was innovative in his administration—he had to be to stay afloat. A perpetual problem had always been the inability to place those children born with congenital syphilis into foster homes. The doctor emptied out an entire section of the orphanage after deducing the simple solution of placement of these unfortunate castoffs with women who had already become immune by contracting the disease at some prior time in their lives. Faced with the high rate of stillborn deliveries among young unwed mothers under public relief, he instituted programs of prenatal care and instruction that soon contributed to a sharp drop in infant mortality to the level of the rest of the population. In so doing, of course, he further increased his own work load by augmenting the number of homeless children. Having watched many of the young women in his charge succumb to the complications of abortions induced at the hands of the famed local "angelmakers," Manchot declared war on the practice and drove most of them out of the business and out of town. But most of all, the doctor simply enjoyed making his rounds at the orphanage where he greeted "Heine" and "Anne," as he called them all, regardless of their real names.

Dr. Manchot soon realized that the small city orphanage would be inadequate in a short time. Working with his colleague, Dr. Peterson, he began to persuade city officials of the need for a much larger children's home. Selecting a site on Winterhuder Weg around the corner from his teenage school of natural sciences and near the existing orphans' home, he drew up careful blueprints that included his own novel design of large balconies in the rear where the youngsters could play and enjoy the sunshine (Figure 28). The war delayed construction until 1922 when dedication was none too soon for the hundreds waiting to move in.[26] This event was perhaps the zenith of Manchot's clinical career.

While Dr. Gillies and his associates were forging surgical history across the Channel during the Great War, Dr. Manchot's thoughts were far removed from cutaneous arterial territories. In addition to all of his public responsibilities and his busy private practice, he was assigned additional duties at a nearby hospital for the care of sick and wounded troops sent back from the front. His 50 years catching up with him, Manchot's health started to fail under the strain. At war's end, his infinite vigor had discovered limits as he was forced to choose among his previous overlapping professional activities. He gave up his insurance practice, keeping the much smaller group of private patients he had cultivated, but continued without respite in his public responsibilities with the children. Further, as the 1920s brought him into professional competition with young physicians with formal training in pediatrics, he found even his small private practice beginning to dwindle. In response, he created and taught a course during the 1920s on health care at the Social Pedagogic School in Hamburg. His thoughts and experiences on the subject were liberally summarized in a 73-page essay published in 1930.[27]

Dr. Manchot also struggled with the social problem of alcoholism in the community. Writing earlier in his career,[28] he had argued forcefully that the local incidence of delirium tremens could be significantly lowered if only the municipal authorities would reduce the number of permits for public houses allowed in various regions of the city. On another occasion, Manchot's direct advice to one of his patients to leave her alcoholic husband resulted in her becoming an independently successful insurance dealer. But this did not mean that he was an advocate of personal abstinence or public prohibition. To the contrary, his education at the Johanneum had given him respect for the classical concept of moderation in recreation as well as other pursuits, and he indeed enjoyed social libation. When in the company of his old medical school friend Dr. Ris, an avid teetotaler, Manchot always concealed a pint of his favorite Kirsch— for the digestion of fat, it was said.

VI

Ferne Berge seh ich glühen!
Unruhvoller Wandersinn!
Morgen will ich weiterziehen,
Weiss der Teufel, wohin?

Ja, ich will mich nun bereiten,
Will—was hält mich nur zurück?
Nichts wie dumme Kleinigkeiten!
Zum Exempel, Dein Blick!
W. Busch

26. The building stands undamaged and is in continued use today as a day-care center.

27. Manchot, C. Die Gesundheitsflege in Erziehungsanstalten Halle. a S Marhold, 1930.

28. Manchot, C. Das Delirium tremens und die Anwendung der Bedürfnisfrage bei der Erteilung von Schankkonzessionen in Hamburg. O. Meissner, Hamburg, 1903.

I see distant mountains glow!
Restless urge to roam!
Tomorrow I will travel on,
Devil knows whereto?

Yes, I will get ready,
I will—what is holding me back?
Only silly trifles!
As for example, your glance!

Privately, Manchot was a bit Puckish. He was a light-spirited, good-natured, pipe-puffing iconoclast who kept a healthy sense of priorities about his work, his family, and his private activities. His religious childhood provided him a generous repertoire of Biblical quotations with which he could tease his wife, disarm an angry patient, or charm an audience. There was seldom an occasion when he could not quote a scriptural passage to illuminate his position. At the podium he invariably adopted the "singsongy" cadence of an old-fashioned rural preacher, perhaps parodying the memories of his father or wishing, by chance, to elicit as successful an audience response.

Second only to the Scriptures was Manchot's deep fondness for the works of the satirist Wilhelm Busch whose verses have punctuated this narrative. Although not widely appreciated in English-speaking countries nor often taken seriously by the literary world, Busch's illustrated stories have been, since the turn of the century, as fondly familiar to German schoolchildren as "Casey at the Bat" has been to their American counterparts. That Manchot greatly enjoyed Busch is hardly surprising in view of their similar sense of the absurd. There was something profound that Manchot found in Busch's humor—a notion that, to paraphrase the satirist himself, acknowledged that beneath the folly of every human interaction, the stark truth lay hiding in the shadow.[29]

It is not surprising that such a broadly educated and socially concerned man should also harbor strong political sentiments. Despite his compassion and efforts on behalf of the underprivileged, Manchot was no socialist and indeed despised all the political liaisons necessitated by his community responsibilities. There was no equivocation, however, about the central feature of his political beliefs—quintessential German nationalism. He shared the pride of his profession, his culture, and the achievements of his countrymen above all other considerations. And like so many others, he felt humiliated by the punitive sanctions of the Treaty of Versailles as once when seen shaking his fists at the sky, tears streaking his cheeks, watching a great Zeppelin circling west across the Channel as part of a war reparations payment. There is probably no question how Manchot would have resolved the inherent tension between his rightist, nationalist passions and his more compassionate and pastoral pursuits had he lived 6 months longer to see the advent of the Third Reich.

Emmi Cropp Manchot was much more than the good sandwich-maker of her husband's early vision. Her bright sense of humor restored his spirits time and again during the difficult war years and thereafter. She was strongest where he was weak and could keep him on track when his own rudder went momentarily kaput. When he failed to hear the phone ringing during the night, she would assure the anxious

29. See the excellent English translations and critique by Dieter P. Lotze, *Wilhelm Busch.* Twayne Publishers, Boston, 1979.

Figure 29. The Manchots' five children: Anne, Fritz, Margarite, Elisabeth, and Gertrud (the youngest) (courtesy of Gertrud Manchot).

parent on the other end that he was on his way and would dump him out of bed to see that the promise was kept. Manchot was a well-dressed gentleman only because Emmi outfitted him and maintained his wardrobe in fitting condition; he was himself so unconcerned over his appearance that he could barely pick out his own clothes.

Carl and Emmi bore five children between 1898 and 1912, all but one of whom are yet living (Figure 29). The children, close to both parents, entered productive careers as might have been expected from the family traditions of both parents. Carl and Emmi saw that all of the children received as much education as possible, particularly since Emmi was anxious that the four girls get the benefits she herself had missed because of her father's death.

In 1916, the Manchots purchased a stately, three-story, twelve-room home in a quiet neighborhood of the Uhlenhorst district where the center of family activities would be for the remainder of the doctor's life (Figure 30). There Carl regularly saw his patients in the spacious front room study with the large windows overlooking the street. The three servants increased the family's leisure time and permitted the Manchots to entertain guests regularly. The children knew they could bring their schoolmates home for dinner at any time, a frequent occurrence because of the fondness all the youngsters felt toward the good-humored doctor. These were the prosperous days of family picnics, trips to the theater to see marionette shows, and walks in the forest.

But the prosperity was not to last.

Figure 30.   The Manchot family home in Hamburg.

———→ VII ←———

Ach, wie oft kommt uns zu Ohren,
Dass ein Mensch was Böses tat,
Was man sehr begreiflich findet,
Wenn man etwas Bildung hat.
W. BUSCH

Oh how often we must hear
That someone did an evil deed
Which we find quite understandable
If we have a little education.

In August of 1914, the Reichstag had authorized the borrowing of 5 billion marks to fund the war and had simultaneously suspended the regulations regarding loan collateral.[30] This decision to finance the war by borrowing rather than by taxing was the seed for what would happen to the Manchot family and millions of their countrymen in the 1920s. The economic control exerted by the Empire during the war was so tightly managed that few were aware of the precarious credit picture. When this control vanished after the armistice, confidence in the mark crumbled. On top of this, the vindictive demands for war reparations by the Allies were dictated by the 1919 Treaty of Versailles "for all the loss and damage to which the Allied and associated governments and their nationals had been subjected." To some of the French and British this meant the inclusion of all pension payments to Allied combatants, the cost of the occupation of the Rhineland, and even the ultimate destruction of Germany as a nation. With growing pressure from American lenders for repayment of their own war loans, the French and British of the Reparations Commission finally settled on an absurd total bill to Germany of 132 billion gold marks.

The central problem was the method by which the Allies could collect the reparations. Germany's gold reserves totaled only 2.4 billion marks at the end of the war, and both labor services and manufactured goods were opposed as *in kind* payments by the Anglo-French labor unions and business community, respectively. The crippling of German commerce by selective tariffs and by the Allied confiscation of the merchant fleet and rolling stock further depressed the national income.

The only solution left to the hapless, deficit-ridden Weimar government was "floating debts" (printing paper money). When the Austrian archduke had been assassinated by a Serbian fanatic in June, 1914, four marks were the equivalent of one American dollar. By the time of the armistice, that figure had fallen to 16 marks to the dollar, and by January of 1922 it had further deteriorated to 190. At the onset of the crippling general strike 12 months later that followed France's occupation of the Ruhr, the exchange had plummeted to 18,000 marks to the dollar. Multiplying over and over again with every day's new quotation, the exchange rate reached 6 figures in July, 7 figures in August, 8 in September, 11 in October, and finally went into orbit at 4 trillion marks to the dollar on November 15, 1923. By that time there were "1,783 presses running day and night to print Reichsbank notes, which had to be transported to banks of issue in large straw crates carried by armies of porters."[31]

⸺⸺ VIII ⸺⸺

Ängstlich schwitzend muss er sitzen,
Fort ist seine Seelenruh,
Und vergeblich an den Zitzen
Zupft er seine magre Kuh.
W. BUSCH

Sweating anxiously, he has to sit there,
Gone is his peace of mind,
And in vain he is plucking
On the teats of his skinny cow.

30. Craig *Op. cit.* See the excellent description of postwar Germany, pp. 435–480.
31. *Ibid.,* p. 450.

The social consequences of the inflation touched the Manchots as well as every other German family. A turnip cost Emmi Manchot billions of marks. The worker living next door, earning a wage of 200 billion marks a month, would get paid twice a day so that he could take half an hour off to run to the store with a basket of currency before the next quotation would halve his earning power. With the rise of the enormous black market, the Manchots had to select which pieces of jewelry or furniture to barter in exchange for vegetables and meat. Having always followed the Protestant standards of frugality and thrift, Emmi and Carl saw the value of their life's financial savings wiped out in a short time. Entertaining of friends and children's playmates was sharply curtailed. The house itself was refinanced and the servants were dismissed. With the older children leaving to seek their fortunes, the vacant rooms were rented to paying guests, all from "proper" social stations, to supplement the family income.

Dr. Manchot witnessed a tripling of the relief rolls from 1913 to 1923. Death rates in cities such as Hamburg increased by over 10 percent between 1921 and 1923 as the financial realities produced their health and nutritional consequences. Only big businesses, borrowing against their hard collateral and paying off in devalued currency, could expand and prosper, using their increasing political clout to prevent the increased taxation that might have stabilized the economy and the government.

Particularly hard hit were the children, making Dr. Manchot's professional role all the more arduous. Public resources to provide meals for the children of the orphanage were grossly insufficient. The poor nutrition of the elementary school children made them listless in the classroom. Fully a quarter of the youngsters in grammar school were substantially underweight and stunted in height by 1923. Dr. Manchot saw a sharp rise in tuberculosis and rickets among his patients. Hamburg's newly constructed children's home in 1922 was faced with the full impact of the national disaster, and its director, though prescient in foreseeing its need, was to be nearly consumed by the consequences.

Providing for the children's education became a particular challenge for Carl and Emmi, especially in view of the revolutionary change in the status of German women with the war. Four girls and a boy would have evoked but a modest expenditure in tuition costs in the nineteenth century when most of the professions were closed to women and secondary education available only in private finishing schools with curricula limited to music, arts, some German classical literature, conventional ethics, and conversational arts. The crushing demand for services during the War had been sufficiently liberating that a legacy of 11 million women had become fulltime members of the labor force by the 1920s. Suffrage, educational opportunities, business openings, and degrees in the professions soon followed. The change in social position was no better illustrated than in the Manchot family, where Carl's mother-in-law had literally mourned the final half-century of her life in black clothing in tribute to her young husband's death; in contrast, Carl's three youngest daughters successfully pursued the professions of education, medicine, and accounting. With most of his private practice gone and with no other source of supplemental income, Manchot was crushed when his two youngest daughters had to work part-time to put themselves through school, something he himself had never had to do. The burden was eased a bit when daughter Elizabeth, with M.D. in hand, took her clinical training in Hamburg and joined her father's practice at the orphans' home for a year.

For consolation Manchot became active in the local medical society and in a prestigious literary group known as the Veronica Society. Here he bantered and debated

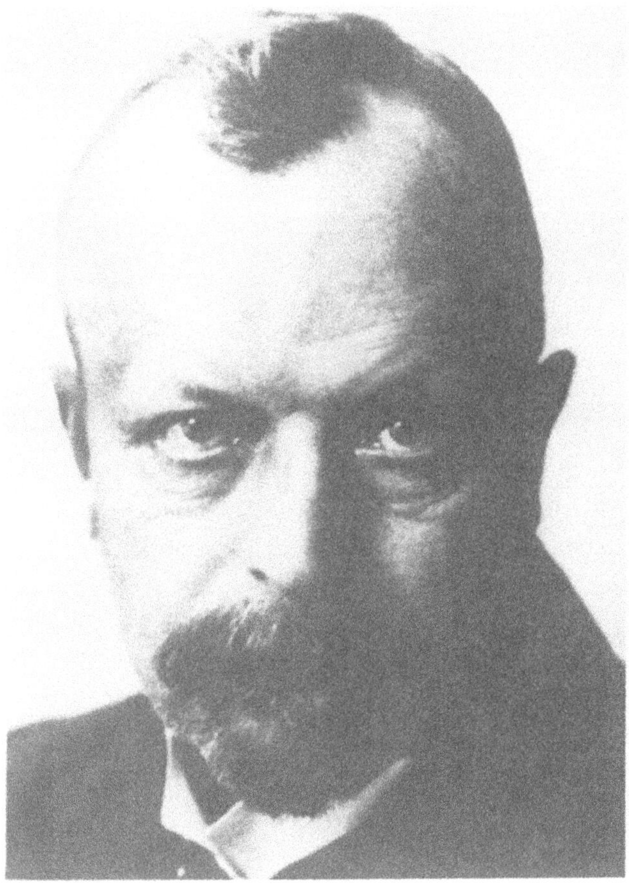

Figure 31.   Dr. Carl Manchot during the Great War (courtesy
of Gertrud Manchot).

gentlemen of like disposition well into his senior years over matters of theology, politics,
and the virtues of German culture (Figure 31).

In September of 1930 (the same year Dr. Gillies was knighted by King George
V for his monumental wartime service), Dr. Manchot enlisted as second ship doctor
on the oceanliner *Deutschland* for a 3-week, round-trip voyage to New York. At the
age of 64, he was badly in need of a respite from his still busy schedule. As he noted
in his personal diary of the journey, "For one beaten down by the struggles of life
nothing is more rejuvenating than a trip to sea." Convinced that ocean air had pharmaco-
logic benefit for his high blood pressure and with reference to supporting opinion of
Bismarck's own physician, Dr. Schwenniger, he often stood alone on the promenade
deck, breathing deeply as he watched the flying fish.

He was at once overwhelmed by New York's grand scale of chaotic order and
its soulless pursuit of the dollar. He admired the discipline of the inhabitants in their
frenzied but methodical negotiation of the streets in and out of motor cars (Figure
32). He was fascinated and a bit appalled by the patchwork of ethnic neighborhoods
and skeptical, at best, that the melting pot myth would ever be realized. He fretted

Figure 32.   New York City in the 1930s.

at the untidiness of the streets, the painted vanity of the women, and the wastefulness
of overconsumption: "A single edition of the *New York Times* delivers enough paper
to wrap six to eight infants." From the deck of a tour boat on the Hudson River, he
particularly admired the great spire of Columbia-Presbyterian Medical Center.

But envy never appeared. The standard against which this city was to be judged
was that of his homeland, and there was no contest. The trees in Central Park were
no match for their German counterparts, although credit was due for their ability to
survive in a "gasoline atmosphere." New Yorkers practiced the unappetizing custom
of gum chewing rather than the more dignified pipe smoking. They radiated a certain
uniformity in their clothing, their slenderness, their beardlessness, and in their straight-
parted hair: "Fat people with beards were right away suspect as Germans." American
jazz was "for the lower drives of mankind but has nothing in common with what the
German calls music." Movies seemed to substitute horribly bad taste and sentimentality
for redeeming values of the soul. But the temples of Wall Street carried the deepest
significance for Manchot since so much of his country's and his own financial misfortunes
of late years had had their roots there. Indeed, his diary here disdainfully called forth
St. Matthew's question of the profit in gaining the world at the cost of one's soul. In
concluding, Manchot expanded the customary cliche by expressing gratitude for having
seen the City: ". . . but I would not like to have to live in New York or to be buried
or cremated there."

Figure 33.   Dr. and Mrs. Carl Manchot in the early 1930s (courtesy of Gertrud Manchot).

Early on the return voyage, he heard an infant's cry while walking below decks in the crew quarters. Investigating the source, he found a newborn stowaway who, it was hastily explained, was being transported back to Hamburg by a cook to be raised by an aunt and uncle. A bit skeptical, Dr. Manchot moved the baby into first-class accommodations under his personal supervision and saw that the child received the doting attention of the most prominent passengers for the remainder of the trip. Having announced that the child would be well placed in a good adoptive home if the story about the relatives were untrue, he must still have felt a little relieved when the couple were indeed found waiting eagerly at the pier on disembarkation.

In July 1932, Carl took Emmi back to his beloved Switzerland for a vacation in the vicinity of Lausanne (Figure 33). Again, he sought his wildflowers on the steep slopes at Gryon-sur-Bex. A rapid climb of 4500 feet on one expedition gave him some shortness of breath for which he retired early the same afternoon. He never awoke, falling victim perhaps to the cardiac irregularities of altitude sickness. After 66 years as a German, Manchot had come back to the land of his birth to die.

Shortly before departing on his last vacation, he had eulogized a departed friend with the verses of his patron Wilhelm Busch which would soon also come true for him:

Wohl dem, der ohne Grauen,
In Liebe treu bewährt,
Zu jenen dunklen Auen
Getrost hinüberfährt.

Zwei Blinde, müd vom Wandern,
Sah ich am Ufer stehn,
Der eine sprach zum andern:
Leb wohl, auf Wiedersehn.

Happy he, who without dread,
Proven in faithful love,
In confidence crosses over
To those dark regions.

I saw, standing on the river's bank,
Two blind men, tired of wandering;
Said one to the other:
Farewell, until we see each other again.

And like the two blind men, oblivious of one another, Carl Manchot and the Plastic Surgeon were indeed to see each other, but not for about three more decades. As if to mark his passing, there would be held within three months a gathering of doctors in an institution on Fifth Avenue ("the most beautiful street in New York," said Manchot's diary)—the first meeting of the American Society of Plastic and Reconstructive Surgeons.

—— IX ——

It is neither fair nor accurate to describe Manchot's book as "lost" for 85 years. A close look at the intervening history of plastic and reconstructive surgery would instead show that although the Gillies flap mainstream was preoccupied with its own set of rules, many creative surgeons within the specialty were indeed intrigued with alternative "arterialized" designs of flap vascularity. While a comprehensive historical review of twentieth century plastic surgery is not feasible, it seems appropriate to cite a few such pioneers, whose independent contributions would ultimately crescendo into the great arterialized flap revolution of the 1970s. Manchot's monograph was certainly known to some, but others pursued arterialized flap concepts without its benefit. As might be expected, the list is headed by two Germans whose scalpels restored the maimed faces and limbs passing eastward out of the same trenches of the Great War.

Jacques Joseph's life span was nearly congruent with Manchot's, although they almost certainly never met (Figure 34). Joseph was not only familiar with Manchot's book, but he keyed at least one reconstructive innovation to it. He refers to the anterior chest diagram (Plates 3 and 4) in his 1931 book, *Nasenplastik und Sonstige Gesichtsplastik,*[32] as anatomic justification for his medially based deltopectoral flap in anterior neck reconstruction (Figure 35). Although Joseph is primarily remembered for his important pioneering work on rhinoplasty, he was an innovator in many other areas of reconstructive surgery as well. A dignified and deeply respected surgeon, Joseph nevertheless lacked that extroverted personality that helped account for Gillies' personal dominance and the dominance of his methods in the English-speaking plastic surgery world.[33]

32. Leipzig: Kabitzch, 1931, p. 673.
33. See Natvig, P. Some aspects of the character and personality of Jacques Joseph, *Plast. Reconstr. Surg. 47:*452, 1971; and Safian, J. Personal recollections of Professor Jacques Joseph, *Plast. Reconstr. Surg. 46:*175, 1971.

Figure 34.   Dr. Jacques Joseph (1865–1934), plastic surgeon of Berlin. (Reproduced with permission from Natvig P. *Jacques Joseph, Surgical Sculptor.* W. B. Saunders Co., Philadelphia, 1982.)

Johannes Friederich Esser, a dozen years Manchot's and Joseph's junior and originally from Holland, offered his services to the Austrian government in 1914 and further pursued what he termed "Structive Surgery" in Berlin after the armistice (Figure 36). Esser developed considerable innovative skill in transferring arterialized "island" flaps on the face and trunk during the war and argued his minority case outspokenly against the Gillies "random flap" tradition thereafter. His prescience for the flap revolution of the 1970s is startlingly revealed in a 1934 article as he summarized his earlier work: "The name 'Artery Flaps' is added to establish the place of the 'Biological Flaps' as the arteries are easily (sic) to find and feel and as *their territory indicates the form and the size of the flaps.*"[34] (italics mine.) Esser's early descriptions of island flaps based on the intercostal arteries for trunk reconstruction and of the epigastric artery for genital

34. Esser, J. F. S. Biological or artery flaps: General observations and techniques. *Revue de Chirurgie Plastique,* January, 1934, pp. 275–287.

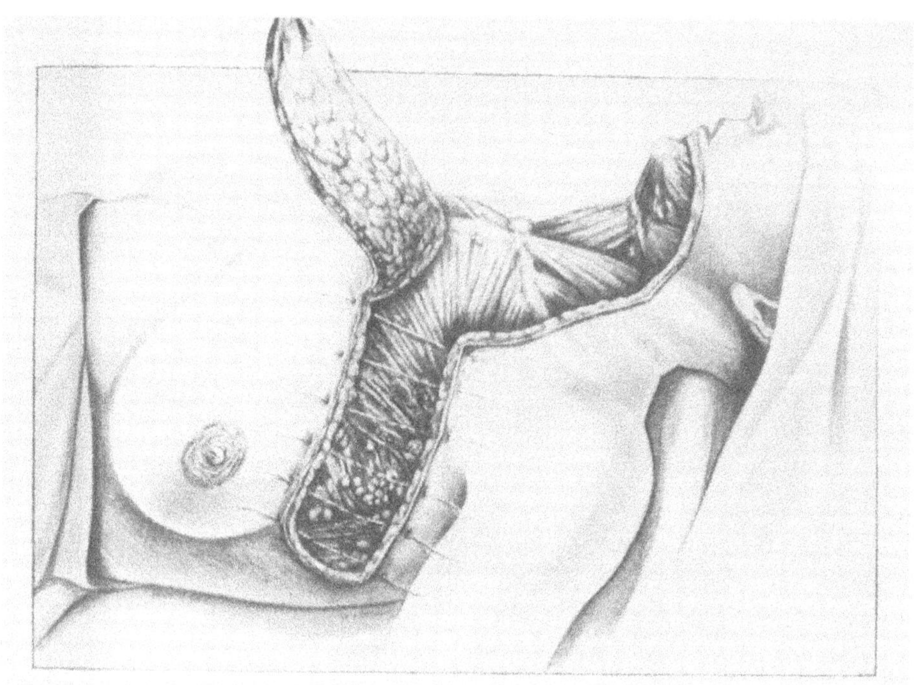

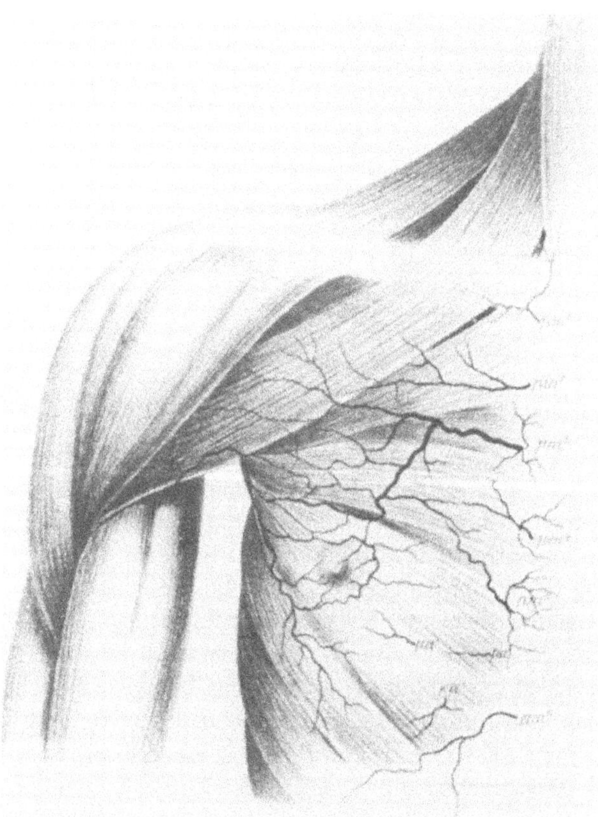

Figure 35 *Above:* Joseph's method of neck burn scar reconstruction with transverse thoracoepigastric flap. *Left:* Reproduction of Manchot anterior chest cutaneous arterial anatomy diagram by Joseph to justify unusual length of the flap. (From *Nasenplastik und Sonstige Gesichtsplastik.* Kabitzch, Leipzig, 1931, pp. 673 and 817.)

Figure 36. Dr. Johannes Friedrich Esser (1877–1946) of Berlin.

Figure 37. Dr. Jerome Webster (1888–1974) of New York (courtesy of Webster Library, Columbia University).

and anterior thigh reconstruction further placed him ahead of his contemporaries. Nevertheless, Esser struggled nearly alone in his enthusiasm for his methods, and his dreamed-for "Esser Institute for Structive Surgery" never materialized.[35] Although closely admired by such prominent plastic surgeons as Gustave Aufricht and Jacques Maliniac and, despite the warm reception given him at a presentation before the American Society of Plastic and Reconstructive Surgeons, Esser's flap techniques persisted merely as interesting curiosities to that huge majority not personally trained to perform them. Not a single reference to Manchot's work can be found in Esser's prolific writings, however. One can only wonder what surgical barriers might have been breached by this perceptive surgeon if he could have but glimpsed the panoply of arterial territories waiting to be exploited.

Across the Atlantic was Dr. Jerome Webster, working in the tall tower of Columbia-Presbyterian Medical Center that Manchot had so admired from a boat on the Hudson River on his 1930 journey to New York (Figure 37). Like Dr. Joseph, Webster was not only familiar with Manchot's book but used it specifically in his clinical work. In particular, the large thoracoabdominal flap that bears his name was keyed precisely to incorporation of the four contributing arteries described by Manchot for the region (Figure 38).[36]

35. See Maliniac, Jacques W. Johannes Frederich Esser (1877–1946), *Plast. Reconstr. Surg.* 2:174–176, 1947.

36. Webster, J. P. Thoraco-epigastric tubed pedicles, *Surg. Clin N. Am.* 17:145–184, 1937 (pp. 153–156).

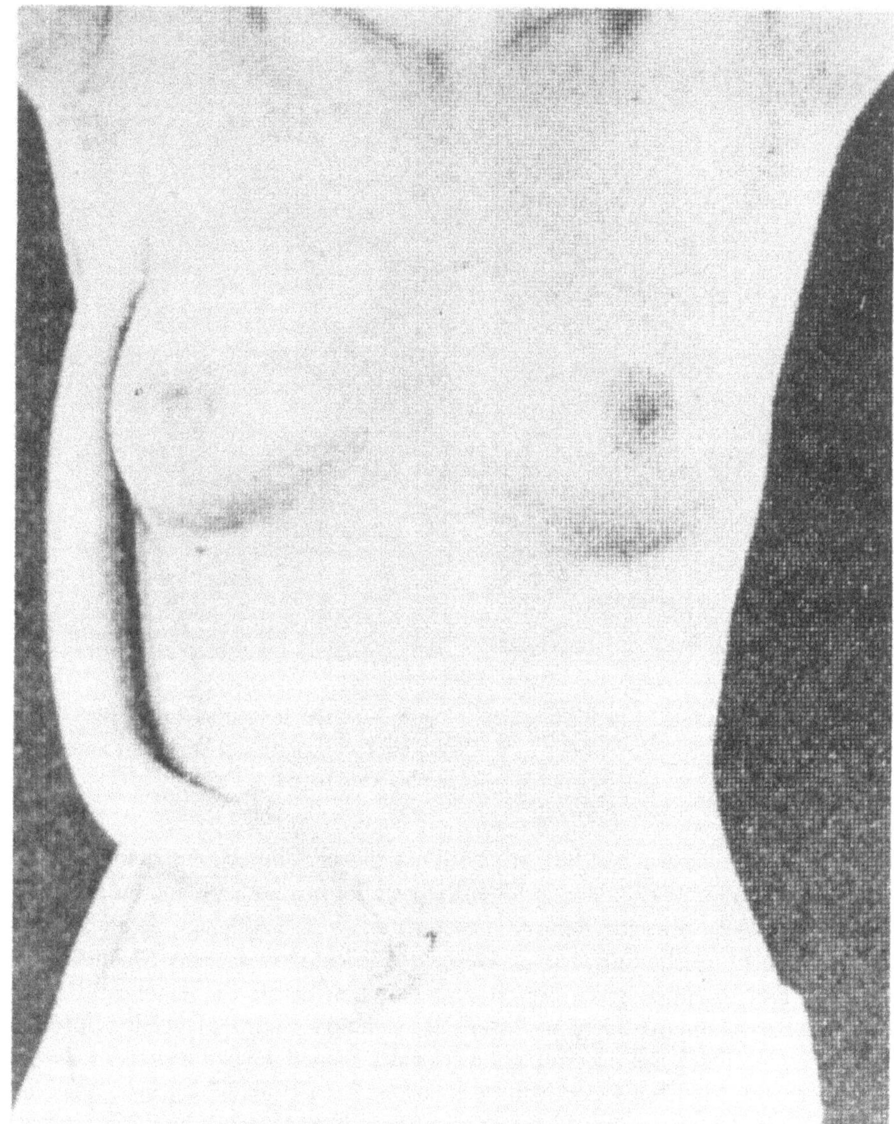

Figure 38. *Above:* Thoracoabdominal flap designed by Webster. *Opposite Page:* Reproduction by Webster of Manchot anterior trunk cutaneous arterial anatomy diagram. Webster cited the four contributing arteries to his flap based on Manchot's work. (From Webster, J. P. Thoraco-epigastric tubed pedicles. *Surg. Clin. N. Am. 17:*145–184, 1937.)

Iginio Tansini was a versatile and innovative Italian surgeon who precociously developed an important arterialized flap in 1906.[37] Tansini experimented with axillary-based skin flaps from the back for reconstruction of the anterior chest skin following mastectomy. When distal necrosis consistently occurred, he consulted the cadavers at the Institute of Anatomy at Pavia and discovered the importance of the arterial intercon-

37. See the excellent discussion by G. P. Maxwell: Iginio Tansini and the origin of the latissimus dorsi musculocutaneous flap. *Plast. Reconstr. Surg. 65:*686, 1978.

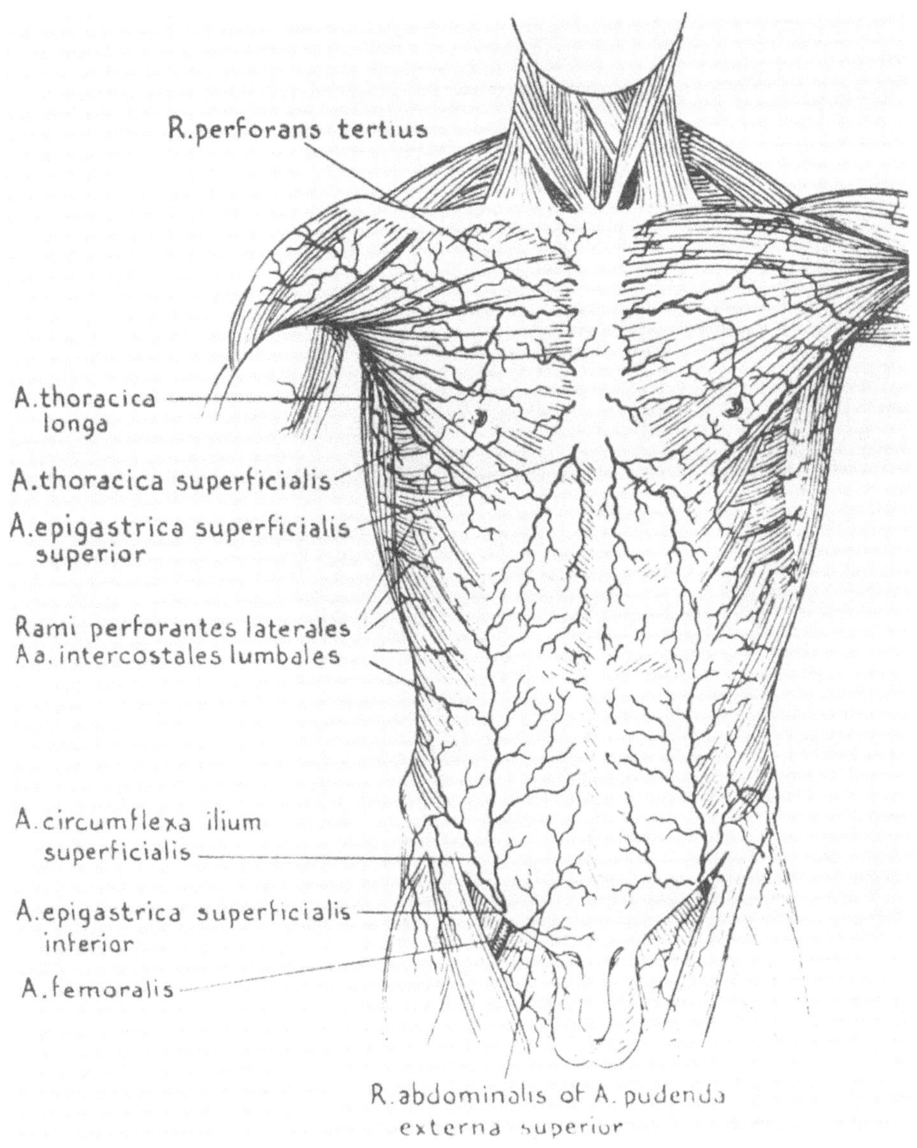

nections joining the major regional vessels, the latissimus dorsi muscle, and the overlying skin. He therefore modified the flap procedure to include not only the skin but a large portion of the latissimus dorsi muscle as well. This modification resulted in a flap that invariably survived in its entirety. This reliable method of breast skin replacement, quite popular in Europe from 1910 to 1920, fell into disfavor with the rising hegemony of Dr. William Halsted whose unwavering advocacy of simple closure of the chest wall with or without skin graft dictated the standard of care for generations. Tansini's clever method was all but forgotten by the time of the flap revolution of the mid-1970s. With some modification, the Tansini method has now become a procedure of choice in secondary reconstruction of the female breast and is one of the most useful legacies of the "arterialized" flap age. That a clinical surgeon working independently outside

Figure 39.   Dr. Vilray Blair (1871–1955) of St. Louis.

the mainstream continental tradition should have followed a logical investigative sequence through the anatomy laboratory to the operating table when so many reconstructive surgical specialists did not is indeed remarkable. That such a superior method should have been permitted to fall passively into obsolescence because of the political dominance of another surgeon is, as in the case of Esser, an unhappy comment on the profession.

Dr. Vilray Blair was a remarkable St. Louis surgeon whose 1912 book, *Surgery and Diseases of the Mouth and Jaws*, brought him prompt acclaim in the then rudimentary field of plastic and maxillofacial surgery (Figure 39). He became, alongside Gillies, an important figure in the reconstructive wartime hospital units of England and was undoubtedly the most important figure in establishing plastic surgery as a specialty in the United States. Well before the war, however, Dr. Blair had recognized the value of including muscle in some flap units, even permitting the transport of bone as well. In one notable instance:

> . . . in a rapidly growing carcinoma which involved half the jaw, the submaxillary region, and the lower part of the face, the patient elected to take his chances on an operation. The nature of the growth demanded that this be done at a single sitting. Before invading

the mouth tumor, we freed the flap of tissue which extended from the level of the mastoid process to 10 cm. below the clavicle, having its attachment above. This flap included the sternomastoid muscle and the middle 8 cm. of the clavicle. After removing the gland-bearing tissue of the neck, the borders of the remaining defect were drawn together. Then the tumor, including three-fourths of the body of the jaw, the lower half of the cheek, and the lower lip, and the tissues of the submaxillary region, was removed in one block. The infraclavicular portion of the flap was turned upward and around the section of clavicle. The flap was brought forward to replace the defect in the face. The clavicular section being fastened to the cut ends of the jaw-bone and the infraclavicular portion of the flap being used to replace the deficient buccal mucosa, left a submaxillary defect which was packed with gauze. Mechanically this worked out fairly well, and had the patient survived, we believe the vitality of the bone flap would have been insured, from the fact that it was surrounded by well-nourished tissues which are normally attached to it. Unfortunately, the patient did not long survive the operation.[38]

The clinical vanguards of the arterialized flap reawakening were Mr. Ian Mc-Gregor's forehead flap and Dr. Vahram Bakamjian's medially based deltopectoral flap, the latter a rediscovery of the Joseph procedure described above. Bakamjian had conceived the concept of the inverted flap for pharyngoesophageal reconstruction in response to a vexing predicament:

> I had let the resident go to the operating room by himself to do a laryngectomy. We thought it was going to be a simple laryngectomy and primary closure, so I did not go to the operating room. He went, and we had misdiagnosed the case. It was more in the pharynx than the larynx and he had to take the whole pharynx circumferentially with the larynx. Then he called me up and said, "What do we do now?"[39]

Solving this immediate problem with a tubed chest flap based laterally on the shoulder, the stage was set for the critical flash of insight which would finally mesh with the Manchot diagrams:

> So the next time around . . . in a similar case, I had to resect the whole thing up much higher. To make it reach high up, we would have to base it medially. I was taught not to base flaps on the midline medially. Then I stopped. I think it came as a flash—that why shouldn't we base it medially? The internal mammary artery is there, and it should work.[40]

It did.

Meanwhile, events were snowballing in at least two other areas. These included the improving technology in surgery of the ultrasmall, some interesting physiologic studies on muscle, and some revolutionary experimental work by an Oxford physiologist.

Stuart Milton (Figure 40) was investigating flaps for a Ph.D. dissertation in the laboratories of the Nuffield Department of Surgery of the Radcliffe Infirmary in Oxford, England, in the mid-1960s when he started to get some "wrong" answers. Raising a variety of paired vertical flaps on the flanks of pigs, Milton found that flaps

38. Blair, V. P. *Surgery and Diseases of the Mouth and Jaws*, second edition. C. V. Mosby Co., St. Louis, 1916, p. 383.

39. Personal communication.

40. Personal communication.

Figure 40.   Dr. Stuart Milton (1935–1971) of Oxford, England
(courtesy of Bert Myers, M.D.).

made under the same conditions of blood supply would all survive to the same length
regardless of their width, a major departure from a fundamental tenet of traditional
plastic surgery dogma. He further concluded that the only effect of decreasing the
width of a flap was to reduce the chance of the pedicle containing a large vessel.[41]
Finally, he cut the skin completely at the base of the flaps to reduce the skin portion
to an "island" and discovered that most still survived to the same length without any
cutaneous attachments at all. In fact, those "island" flaps containing a segmental artery
(the equivalent of those vessels described by Manchot) would survive to a *50 percent
greater length* than those flaps with cutaneous continuity which had no specific vessel
in the base. At last, the proper functional question had been addressed in the laboratory.
Plastic surgeons would no longer be shackled by the honored but mistaken traditions
of Milton's fellow England-based surgical pioneer who had died seven years previously
at the age of 76.

41. Milton, S. Pedicled skin flaps: The fallacy of the length:width ratio. *Br. J. Surg. 57:*502,
1970.

Milton was fascinated with Manchot and recognized at once the remarkable achievement in the German's book. "Those who have not seen the segmental vessels in man during paramedian incisions or radical mastectomies might dismiss the monograph of Manchot (1889) for lack of clear experimental details and evidence," Milton wrote in 1967, "but it is a remarkable coincidence that his description of the largest vessel of the ventral chain (second intercostal) is the one included in the versatile delto-pectoral flap . . . and the one which underlies the warmest area of the ventral chest wall."[42] Milton arranged for Brian Ross, a colleague at Oxford, to do an English translation of Manchot's book from which to work. Milton carried this with him in his American travel and in his important collaborative studies with Dr. George Cherry of Ann Arbor and Dr. Bert Myers of New Orleans. He circulated the copy to others involved in the early arterialized flap revolution, including Dr. Rollin Daniel and Dr. William Grabb. Milton was a complex individual who, while recognizing that his crucial investigations would probably revolutionize plastic surgery, became periodically despondent over the early nonacceptance of these ideas. In passing, Milton once remarked to Dr. Myers that, should he die unexpectedly, he wished that Myers would dissect his own anterior chest wall at autopsy to see if Manchot was correct in his descriptions. It was indeed during a visit to New Orleans when Stuart Milton's life ended tragically in 1971, following Manchot and Esser in failing to see the widespread fruition of his work. Dr. Myers chose not to carry out Milton's request.[43]

Dr. John McCraw was an orthopedic surgical trainee at Duke University whose NIH-funded residency demanded some research commitment. As part of an ongoing membrane physiology investigation, he was assigned some flux studies on normal and ischemic frog muscle cells. "Island" muscle flaps were created for the study, divided at both origin and insertion, but these muscles surprisingly did not act very ischemic. Switching projects to study Volkmann's ischemic contracture, again Dr. McCraw found "island" muscle flaps to be very hard to "kill," a second research "failure."[44] Through a series of fortuitous circumstances, Dr. McCraw then carried these observations through a remarkable odyssey of clinical innovation in concert with a series of surgical iconoclasts. Starting in junior association with Dr. Louis Vasconez, Dr. Leonard Furlow, and Dr. M. J. Jurkiewicz in Gainesville, Florida, and later with Steve Mathes, Foad Nahai, John Bostwick, and Jurkiewicz in Atlanta, the potential for muscle and combined muscle-skin flaps was realized. Atlanta's Grady Hospital would become to the "myocutaneous" flap what Sidcup had been to the tubed pedicle in 1917. Further advances accrued when a serendipitous throw of Pentagon dice appointed Dr. McCraw as junior officer to Col. David Dibbell at Wilford Hall USAF Hospital in San Antonio in a happy blend of creative unorthodoxy. The resulting variety of myocutaneous flaps, well arterialized on a pedicle of muscle, were found to be transferable about their vascular axes for a remarkable panoply of practical reconstructive tasks. The fact that the muscle could carry the overlying skin with it was an important new corollary to Milton's laboratory studies. McCraw's anatomic dissections of submuscular and musculocutaneous vasculature form an important functional complement to Manchot's work. For his unprecedented contribution, Dr. McCraw received the William Beaumont Award in Medicine from the American Medical Association in 1980.

42. Milton, S. Experimental studies on island flaps: (1) The surviving length. *Plast. Reconstr. Surg. 48*:1971, p. 576.
43. Personal communication.
44. See McCraw, J. B. The recent history of myocutaneous flaps. *Clin. Plast. Surg. 7*:3–7, 1980.

As microsurgical instrumentation improved during the 1960s under the leadership of Dr. Harry Buncke of San Francisco, it soon became clear that very small blood vessels no greater than a millimeter in diameter could be reliably joined together. To plastic surgeons, this promised an exciting alternative to traditional reconstructive surgery, for a flap might conceivably be detached *completely* from its original site and be rejoined where needed by microscopic suture of its vessels. The only remaining task seemed to be in locating a dispensable cutaneous donor area which was reliably supplied by an artery and vein of ample size and of predictable course and distribution. Manchot's book provided the answer.

Over the years several reconstructive surgeons had been struck by the unusual reliability of skin flaps drawing their blood supply from the inguinal region. As already noted, Dr. Jerome Webster and Dr. J. F. S. Esser had consciously planned large arterialized anterior trunk flaps in such fashion that the superficial inferior epigastric arteries would carry them. In 1946, U.S. Army hand surgeons Darrell T. Shaw and Robert L. Payne presented their independently derived experience on the use of this same arterialized flap for reconstruction of the hand in World War II casualties.[45] Ian McGregor and Ian Jackson of Glasgow focused in 1972 on the contiguous superficial circumflex iliac artery and its more laterally directed distribution in their description of the "groin flap."[46] However, it was the dye flush studies of Dr. Ernest Kaplan of Stanford in the late 1960s and the further dissections of Bernard O'Brien, Rollin Daniel, and Ian Taylor, working in Melbourne, who discovered the reciprocity of these two vessels and the ability of either artery to carry the sizable territory of the other.[47] Kaplan and the Melbourne group all worked from copies of the Manchot book in the course of their investigations. In 1969, Buncke and Kaplan successfully executed a microvascular "free transfer" of McGregor's groin flap to the floor of the mouth with full survival, only to see its edges separate and its tiny pedicle twist and strangulate three weeks later.[48] Three years thereafter Dr. Daniel anastomosed the superficial inferior epigastric artery (described and named in Strassburg 83 years earlier) to the posterior tibial artery in a groin flap free transfer to reconstruct an ankle defect. This time the pedicle did not twist, and Manchot's promised legacy was at last fulfilled.

It is futile to try to segregate medical history from the history of contemporary world events. So it is with military, art, literary, and other historical subdisciplines. That the plastic surgeons of World War I and its long aftermath missed the opportunity offered by Carl Manchot's dissections is not the result of scholarly slothfulness. Gillies was separated from Manchot by those very social forces which separated the nations of the world and brought about the Great War in the first place—chauvinistic nationalism, deficient communications, language barriers, and commercial rivalry, the stepchild of

45. Shaw, D. T., and R. L. Payne, Jr. One stage tubed abdominal flaps. *S.G.O. 83*:205, 1946.

46. McGregor, I. A., and I. F. Jackson. The groin flap. *Br. J. Plast. Surg. 25*:3, 1972.

47. Daniel, R. K., and G. I. Taylor. Distant transfer of an island flap by microvascular anastomoses: A clinical technique. *Plast. Reconstr. Surg. 52*:111, 1973.

48. Kaplan, E. N., H. J. Buncke, and D. E. Murray. Distant transfer of cutaneous island flaps in humans by microvascular anastomoses. *Plast. Reconstr. Surg. 52*:301, 1973.

scientific advancement. If the war catalyzed the development of plastic surgery, the forces behind the war detoured and distorted it.

And if the unseeing loyalty of the patriotic foot soldier and of the home front permitted perpetuation of the war, so did this human trait retard the necessary correction of plastic surgical methodology. For it was to be the doughboy spirit of the trenches that would prevail throughout flap surgery for the next half century. Despite the "heavy casualties" in dead and dying flaps with barely perceptible advances across the front, it was still safer to endure the known mire of the trench than to venture over the top into the unknown perils of no-man's land. There at least was the supposition that one's generals possessed strategic information based on reliable intelligence to justify the prevailing surgical tactics.

Thus few mourned the passing of the Gillies flap tradition, which fell in the 1970s like the Maginot Line. To say it all need never have happened would be to deny the currents of history; to overlook the lesson of Manchot would be to deny surgery its brightest future.

Foreword

The following work was accomplished in response to an open competition held by the Kaiser-Wilhelm University Medical School in Strassburg. The assignment was to fill existing gaps in the knowledge of the arterial supply of the skin. The assignment further specified a comparison of the cutaneous arterial and nerve supplies and an investigation of the interrelationships that might exist between arterial distribution and the somitic arrangement, thus leading to an understanding of the basic organization of the skin arteries. The investigation was carried out in the Anatomical Institute in Strassburg. The necessary materials were placed at my disposal very generously by Professor Schwalbe, who also encouraged me to publish this work. I would like to express my thanks to him for this as well as for his frequent suggestions.

I am much indebted to the publishers for their friendly cooperation, especially for the lovely unabridged reproductions of the plates. It seemed to me that the publication of these plates was especially important as a necessary foundation for further expansion in this field. I was able to rise to this challenge only with the faithful help of my friend L. Bamberger. With the assistance of the dissections and sketches I prepared, he carried out with skillful hand all the drawings that are the basis of the plates. These drawings are without exception drawn according to my dissections insofar as the arterial distributions are concerned. To complete several pictures, the outstanding muscle drawings from the works of Gegenbaur and Sappey were used as the background on which to draw in the arterial courses. Mr. A. Schneegans was kind enough to provide the two whole-body sketches.

First Part

Whereas the nutritional relationships of almost all other organ systems have been accurately investigated and described, those of the skin are only partially known. A systematic, coherent investigation is totally unavailable at present. Only the cutaneous arteries of the head, the foot, the hand, and the knee and elbow joints, which catch the eye because of their conspicuously large size, have been accurately described. For the remainder of the body a number of incidental remarks are found scattered in the anatomic literature indicating this or that area of the skin to be the area supplied by a specific artery. But on the whole the investigation of these relationships has been neglected until now. The origin of the cutaneous arteries from the muscle layer, their course in the fascia, and their final arborization in the dermis have been largely overlooked for many areas of the skin.

A comprehensive investigation of the cutaneous arteries must therefore first collect and compare all the remarks scattered throughout the anatomic literature and reject the more dubious of them through careful cross-checking and verification. This investigation must fill in the gaps of the known anatomy and, where possible, provide through dissection the data which are entirely absent.

Special attention will be directed to discovering as accurately as possible the existence of recognizable regularities in the distributional patterns, directions, and arrangements of the arteries.

Finally, the second problem in a comprehensive study of cutaneous arteries will be to postulate an overall organization that the various factors suggest when taken together.

The following work is an attempt to do justice to this dual problem.

Chapter I

Cutaneous Arteries
of the Trunk

Because the arterial supply of the skin of the back and lumbar region is relatively simple, it is useful to begin with its description. We shall exclude the nuchal and shoulder regions, which are more appropriately described with another cutaneous region.

The channels through which the skin of the back receives its arterial blood are known.

Principally, the intercostal, lumbar, and sacral arteries supply the skin of the back through their musculocutaneous branches. The supreme intercostal artery, which supplies blood to the cutaneous region overlying the first two intercostal spaces, and the iliolumbar artery, supplying the area of the fifth lumbar vertebra, complete the cutaneous blood supply of the back. They share with the proper intercostal and lumbar arteries an analogous pattern of musculocutaneous branches.

All of the arteries that supply the whole expanse of the back display a great uniformity of arrangement including the complementary branches arising from the subclavian artery (supreme intercostal artery) and the hypogastric artery (iliolumbar artery).

Directly adjacent to the corresponding vertebra, these arteries[1] deliver a dorsal branch,[2] which penetrates into the musculature of the back. After a short course, each of these dorsal branches divides into a spinal branch and a

1. Compare Henle, pp. 131, 163, 157–158; Sappey, pp. 566–567, 654; Theile, p. 167; Tiedemann, Table IX.

2. In the case of the intercostal arteries the dorsal branch emerges through the opening bordered by the necks of the ribs and the superior costotransverse ligament. Henle, p. 158; Gegenbaur, p. 699.

musculocutaneous branch. Only the latter is important for the present investigation. It[3] penetrates to the skin surface by way of the back muscles to which it sends branches. Specifically, one row of branches is given off between the layers of the transverse spinal muscle[4] to the area of the vertebral spines (medial branches) and another row between the two divisions of the sacrospinalis muscle passing as far laterally as the cutaneous area of the angle of the ribs (lateral branches). The medial branch[5] penetrates as far as the most superficial muscle layer, thereby passing always closer to the median line. Ultimately, it emerges from the muscle layer directly adjacent to the top of its corresponding lumbar spine, suddenly changes direction, and, coursing laterally, passes into the dermis for its terminal arborization. In addition, the proper intercostal and lumbar arteries[6] give off perforating branches in the posterior and lateral portion of their course that pass anteriorly to the skin of the trunk.

The dorsal branches of the sacral arteries[7] pass through the anterior sacral foramina into the sacral canals. After giving off spinal branches, they pass through the posterior sacral foramina to the posterior ligaments of the sacroiliac joint to the origins of the long back muscles and the gluteus maximus muscle, and ultimately to the skin of the lumbar region.

The arterial supply of the cutaneous areas of the upper intercostal spaces is complicated by the fact that the transverse cervical artery also participates.[8] The terminal branch of this artery (descending branch of Henle, circumflex scapular artery)[9] passes inferiorly at the medial edge of the scapula, enclosed between the rhomboid muscles and the serratus posterior superior muscle. This branch gives off lateral and medial divisions, the lateral supplying the infrascapular region, and the medials distributing themselves between the serratus posterior superior, rhomboid, and trapezius muscles. Individual branches perforate through this muscle and thus establish a relationship to the skin.

This summary of the scattered remarks about cutaneous arteries of the back found in handbooks of anatomy sheds light on the origin of these vessels. However, the courses and relationships of the cutaneous arteries were not taken into consideration until now. On the basis of new investigations oriented from this point of view, the following presentation will attempt to fill in these gaps.

Plate I[10] presents the typical course and relationships of the cutaneous arteries of the back.

3. Henle, p. 158; Theile, p. 167.

4. According to Sappey (p. 567) among others, these medial branches course between the dorsalis longus muscle and the transversospinalis muscle.

5. Sappey, p. 567.

6. Henle, p. 159; Theile, p. 168; Tiedemann, Tables IX and X.

7. Arnold, p. 527. The first dorsal branch is the largest and usually emerges twice.

8. Theile, p. 129.

9. Sappey, p. 623; Henle, p. 134.

10. For special clarity on this drawing, on the right half are presented the cutaneous arteries arising out of the median line, and in the left half, the remaining arteries.

At the points of exit described in detail above, namely the spaces between the lumbar spines, the medial rami[11] of the musculocutaneous branches course to the surface and arborize laterally in the dermis of the back. The superior boundary of this territory that they supply is defined by a line that connects the spine of the first thoracic vertebra with the axis of insertion of the trapezius muscle and the triangular origin of the spine of the scapula. The boundary line then follows the lateral edge of the trapezius muscle inferomedially until it intersects the line defined by the costal angles. It follows this latter line inferiorly to the level of the tenth thoracic vertebra, where it angles more medially toward the lateral border of the lumbodorsal fascia in the lumbar region. Thus it reaches the iliac crest and bends medially to the spine of the fifth lumbar vertebra. In the sacral area there are no medial branches.

These borders are not invariably maintained. They are sometimes expanded and sometimes restricted. When the vessels are especially strongly developed, the vascular territory may be spread beyond the border of the trapezius muscle onto the skin of the infraspinatus fossa.

Restrictions of the territory, when they occur, are evident in two areas.

The first is in the upper line extending from the first thoracic vertebral spine to the scapular spine. The medial branch corresponding to the first intercostal space is often very rudimentary, demonstrated only with great difficulty or entirely lacking. The second medial branch often presents the same appearance as well.

Second, frequently the medial branches in the lumbar region (namely the first through third lumbar interspaces) are very poorly developed or missing completely. In addition, the region of the lowest thoracic vertebrae (nine to twelve) frequently shows a similar situation.[12]

In this area the medial branches spread out specifically in an arrangement that shows unmistakable relationships to the vertebral and rib segments. In individual cases there are some variations. The superior five or six branches distinctively converge largely parallel to the muscle bundles of the trapezius muscle toward their origin in the triangular region of the scapular spine. In Figure 1 the uppermost branch (m_1) goes off to this spot at an acute 60–70-degree angle to the median line. At m_4 this angle is already a right angle and at m_5 it has become obtuse. Even the following branch, m_6, whose principal direction again corresponds to the vertebral segments, sends a strong branch in the direction of the origin of the scapular spine.

Other deviations are a result of the fact that individual arteries that lag behind in development (in Figure 1, m_7 and m_{10}) are replaced by their

---

11. Compare Figure 1 (Plate I).

12. Cases in which the medial branches are lacking to an even greater degree are apparently a result of imperfect injections. At least it was possible for the author, in a case that was particularly striking in this regard, to trace the vessels that were only partially injected through nearly the entire extent of the region.

more strongly developed neighbors (m_8 and m_9). With their richer development, these may then encroach in their distribution on two, three, or more vertebral segments (m_6, for example, in Figure 1). These distributional patterns, however, usually correspond to their respective vertebral segments in the general direction of their course.

The medial branches of the lumbar arteries usually course in a horizontal direction. The first occasionally encroach beyond the territory of the lumbodorsal fascia onto the area overlying the latissimus dorsi muscle. In the lumbar area alone the main trunks may often manifest a highly tortuous course in short S-shaped curves.

All of the medial branches give off numerous acutely angled side branches that anastomose with one another and also with twigs from the lateral branches (and in the superior area in addition with the branches from the transverse cervical artery).[13]

As a rule the median branches are strongest in the midthoracic region (third to seventh thoracic vertebrae). Above and below they are less dominant.

The lateral rami of the musculocutaneous branches (from the dorsal branches of the intercostal arteries) also share in the arterial supply of this circumscribed skin territory. A reciprocal relationship exists between the lateral rami and the medial branches. If the latter are only poorly developed, the lateral rami are more richly developed, and vice versa. In general these comments concerning the medial branches are also true for the lateral branches. They are subject to the same principal laws of arrangement with the same modifications.

Their points of emergence from the muscles are, with the exception of those in the sacral area, not very clearly defined. For the lumbar region, it appears that no generally valid statement can be made. However, this is more readily done when describing the area of the lower thoracic vertebrae, where the exit points in the majority of cases can be connected by a line parallel with the spinal column that divides in half the area between the vertebral spines and the angle of the ribs.

The strongest development of the lateral cutaneous branches is demonstrated in the lumbar region and in the area corresponding to the lower thoracic vertebrae (eight to twelve).[14] In the upper territory, that belonging to the trapezius muscle, they are usually entirely absent. In the sacral area, where the medial branches are missing, the lateral branches are also only rudimentary. The lateral branches pierce the fascia at sites corresponding to the posterior sacral foramina, but only rarely are all arteries present on both sides of the midline. Usually only the first pair is complete. These arteries run in a horizontal direction and occasionally reach the origin of the gluteus maximus muscle.

13. In addition, they stand in anastomotic connection with the corresponding arteries of the other half of the body through tiny branches that come off just before they emerge out of the muscle mass.

14. Compare also Theile, p. 168.

One unusual feature of the lateral branches must still be mentioned: These do not always supply the cutaneous distribution of a vertebral rib segment that corresponds to their origin.

That part of the above-mentioned thoracic territory that is not supplied by the dorsal ramus conforms greatly to this general pattern. As one sees from the handbooks, it is direct branches of the intercostal and lumbar arteries that pierce the muscular layer and supply this territory. We will name them posterior perforating branches to simplify this schema.

Their exit points from the muscular layer cannot be precisely identified by a general description. According to Theile, they usually emerge next to the lateral edges of the iliocostalis muscle to ramify in the latissimus dorsi muscle and the skin. Their number varies from case to case. In an area that in one case may be supplied by a single skin artery, there may be in another case two branches coming from the same intercostal, or a lumbar artery may share in the same function and vice versa. But they then demonstrate the same arrangement in correspondence to the vertebral segments and the same laterally directed course that was described for the medial region. Naturally there may be found here modifications such as those described above, demonstrating the same complementary encroachment of individual arteries in a territory served by other poorly developed ones, a condition that is not at all surprising when one takes into account the rules that govern the distribution of the arteries.

The lateral portion of the back region extends to the area of the lateral trunk wall, with which it is in close anastomotic connection. The term "lateral trunk region" refers to that area[15] whose dorsal boundary is represented by the lateral edge of the latissimus muscle, with a caudal border at the iliac crest and ventral border in the chest area at the lateral edge of the pectoralis major muscle and the lateral edge of the rectus abdominus muscle. A sharply defined boundary is lacking in the lower abdominal area. Therefore, this area lying between the lower boundary of the ribs and the iliac crest will be more appropriately described in connection with the cutaneous arteries of the anterior abdominal wall.

The nourishment of the lateral trunk region, from the lateral border of the latissimus dorsi muscle on,[16] also occurs for the most part via perforating musculocutaneous branches of the intercostal and lumbar arteries. In general, the same rule applies for them as for the skin arteries of the back. However, some unusual features of these lateral perforating branches, as we shall call them, require a somewhat more detailed description. In addition, the arterial arrangement in the upper chest region is further complicated by the substantial participation of the axillary thoracic arteries.

The origins of the lateral perforating branches from the intercostal arteries can be connected by a curve that extends, with dorsally oriented convexity,

15. Compare Figure 2 (Plate I).
16. Compare Figure 2 (Plate I).

around the lateral lower edge of the pectoralis major muscle to the lateral boundary of the external oblique muscle to the twelfth rib. From the fourth, fifth, or sixth intercostal space onward, these cutaneous arteries penetrate the musculature and appear on the surface at the slips of origin of the external oblique muscle. Their origin thus lies in the upper portion of the region from the fifth (sixth) to the eighth intercostal space between the insertion slips of the serratus anterior muscle. In the lower portion (ten to eleven) of the same region they are enclosed by the three costal origin slips of the latissimus dorsi muscle. The artery arising from the ninth intercostal space emerges between the last serratus slip and the first costal latissimus slip.

Directly at its point of exit on the surface of the trunk wall, one of each of these arteries divides into two branches, which diverge in opposite directions.[17] The usually more prominent anterior branches travel anteriorly in a horizontal direction parallel to the ribs and supply via copious ramifications the corresponding territory of skin above the external oblique muscle. The less prominent posterior branches pass dorsally, also parallel to the ribs, and reach the skin overlying the lateral border of the latissimus dorsi muscle, to terminate their ramifications.

The division of the perforating branches into diverging cutaneous branches does not always occur where they emerge from the muscular layer. Specifically, they often ramify while still within the muscle in the upper region (fourth to seventh intercostal spaces), and the cutaneous branches, which, according to their origin, should belong together, then emerge on the surface in two different locations, which may be several centimeters apart from each other. By means of finer secondary branches that travel in the opposite direction of their artery, an anastomosis is usually created between them and the supply of the skin territory lying between the exit points.

In addition to these specific arteries, other lateral perforating branches also emerge from the intercostal arteries to the skin of the lateral trunk wall.

They penetrate in a number that varies from case to case at varying locations on the external oblique muscle and become aligned completely in course and orientation with this muscle. They run anteriorly in the area of the anteriorly oriented and posteriorly in the area of the posteriorly oriented cutaneous branches. In accordance with the known variability and development of the latter in individual cases, they constitute an important supplement to the arterial supply of this region. The variations and modifications of the principal pattern of cutaneous circulation of the back naturally holds true here as well.

In the uppermost regions of the lateral chest wall, penetrating branches of the intercostal arteries[18] also take part in supplying the skin. But they play an extremely minor role in comparison with the branches usually involved here,

---

17. See Figure 2 (Plate I).

18. In Krause's tabular overview of "the blood vessels of the individual organs," the cutaneous branches do not arise from the intercostal arteries. (See older edition of Krause's Anatomy, p. 1172, and see new edition of W. Krause 1880, p. 270.)

those associated with the axillary artery, viz. the long thoracic and subscapular arteries.[19] The arterial territory of the lateral chest wall, primarily that of the cutaneous branches of the axillary artery, can be defined[20] by a triangle whose base is formed by the fourth and fifth ribs, whose apex is the hollow of the axilla, and that is bounded laterally by the lateral edges of the pectoralis major and latissimus dorsi muscles.

It is known that the long thoracic artery and the subscapular arteries, which are brought into the discussion at this point, are exceptionally variable.[21] Whereas they are sometimes represented by numerous branches coming from the axillary artery, they are often consolidated into a very few or even into a single trunk.[22]

The long thoracic artery sometimes consists of two parts[23] and often is reduced in its vascular territory.[24] Frequently it is entirely lacking, and its territory is then supplied by the thoracodorsal branch of the subscapular artery. Alternatively it can be very strongly developed and then may replace the thoracodorsal branch (often, according to Cruveilhier). In other cases the long thoracic artery is a side branch of the subscapular artery (the normal situation, according to Meckel).[25] An accessory long thoracic artery may also originate from the subscapular artery. The subscapular arteries are also subject to many diverse variations.

It is clear that no absolute rules may be set forth for the origins of the cutaneous arteries of this area. The leading textbooks present, with some justification, the following arrangement as the norm. The long thoracic artery[26] descends on the serratus anterior muscle to its anterior edge, at the fifth or sixth intercostal space. Anteriorly and posteriorly, it gives rise to branches that supply the serratus and that anastomose with branches of the intercostal arteries and of the subscapular artery.

It is covered by the pectoralis major muscle in the superior portion of its course and is subcutaneous in the inferior portion.[27] It also sends branches to the breast. The subscapular arteries divide into two or three superior and one more prominent inferior branch, "which courses over these muscles and divides into the circumflex scapular artery and the thoracodorsal artery at the lateral edge of the scapula. The thoracodorsal artery descends along the lateral wall of the thorax to its inferior edge, between the serratus anterior and latissimus dorsi muscles, arborizing in both of them, posterior to the long thoracic artery and anastomosing with it and with the intercostal arteries, and anastomosing

19. Compare Figure 2 (Plate I) and Figures 3 and 4 (Plate II).
20. Compare Figure 2 (Plate I) and Figures 3 and 4 (Plate II).
21. Gegenbaur, p. 609.
22. W. Krause, the varieties of arteries. Henle, p. 266+.
23. Author's observation.
24. Krause, p. 267.
25. This variation was also frequently observed by the author.
26. Sappey, p. 628+; Henle, p. 138; Gegenbaur, p. 689+; Arnold, p. 489+.
27. Sappey, p. 628.

with the descending branch of the transverse cervical artery at the lower corner of the scapula."[28]

The lymphatic glands and skin of the axilla and of this general region receive cutaneous branches from the inferior subscapular artery.

Although the origin of the cutaneous arteries in the upper lateral chest region is exceptionally variable, their more distal branching and terminal ramifications to the skin exhibit a certain uniformity in arrangement.

From the main trunks of the thoracic and subscapular arteries, which intersect the course of the ribs at right or acute angles, many lateral branches arise and run more or less parallel to the ribs in an anastomotic network. A variable number of perforating branches from the intercostal arteries (lateral perforating branches) contribute to this network. This produces a picture that is very similar to that of the cutaneous branches of the intercostal arteries in the inferior region (compare Figure 2, Plate I, Figure 4, Plate II).

As previously mentioned, the arteries of this region anastomose with the cutaneous arteries of the anterior chest.[29] The term *anterior chest region* refers to that skin surface that covers the pectoralis major muscle and the anterior surface of the sternum.

A description of the arteries of the breast cannot be excluded from this presentation. In terms of its arterial supply, the breast appears only as a very richly nourished region of skin that cannot be separated from the surrounding areas. The developmental history further justifies including the arteries of the breast with the arteries of the skin and will therefore be described along with them.

The sources from which this region obtains its arterial blood are known: (1) branches of the anterior intercostal arteries that send perforating musculocutaneous branches into the region; (2) the perforating branches of the internal mammary arteries; and (3) the thoracic arteries. Of this last group, the long thoracic artery (see above) and the thoracoacromial artery are frequently mentioned in the literature. In addition, Arnold, Krause, and Henle[30] mention the participation of the supreme thoracic artery (external supreme thoracic artery of Arnold) in supplying the skin of the breast.

The perforating branches from the anterior intercostal arteries are only of minor importance to the supply of this area.[31] Only in the inferior portion do they appear with any regularity; one or more may be present scattered in the superior portion, but they are usually completely absent. As a rule, they do not appear superior to the fourth rib. Only rarely does one find branches in the third intercostal space. In only a single case could a small skin artery of the second intercostal space be traced with certainty to the lower branch of the second intercostal artery, and this supplied only a limited skin area. Even

28. Henle, p. 139.
29. Compare Figures 3 and 4 (Plate II).
30. Arnold, p. 488; C. FR. TH. Krause, p. 1172; W. Krause, p. 272; Henle, p. 136.
31. Plate II, Figures 3 and 4 pa.

in the inferior region, where these branches are stronger and more numerous, they play only a subordinate role.

In terms of number, exit point, and territory of distribution, they are subject to great deviation from case to case. However, they always exhibit the directional arrangement parallel to the ribs so characteristic of the branches of the intercostal arteries. In his anatomy description, Hyrtl[32] asserted that the arteries of the areola–nipple complex originated exclusively from the fourth intercostal. On his authority this information has been incorporated into almost all handbooks. Henle[33] also accepted this hypothesis without question. Only Krause[34] doubted its validity for the female breast. In the author's investigations, particularly abundant in this area, Hyrtl's assertion was not confirmed in any case, including that of the male breast. On the contrary, it became apparent that the principal supply of the areola–nipple complex may be ascribed to the perforating branches of the internal mammary artery and the thoracic arteries.

The perforating branches of the internal mammary (six to seven in number) exist in a very characteristic relationship for the nutrition of the anterior skin and glands of the breast.

They pass through the corresponding intercostal spaces and divide into muscular branches (for the pectoralis major) and skin branches.[35] The cutaneous branch of the first perforating branch originates between the sternoclavicular joint and the first rib. For the remainder, a single intercostal space corresponds generally to one more or less developed cutaneous branch. Occasionally, however, one finds two of them (compare Figure 4, Plate II at the first intercostal space). They emerge on the surface at the medial edge of the pectoralis major muscle and course laterally in the subcutaneous tissue and skin of the breast. The territory they supply is very extensive. They leave only the inferior and superolateral skin overlying the pectoralis to the thoracic arteries.

A narrow band, corresponding to the lower edge of the central clavicle at the first intercostal space, is supplied by the arteries of the neck. In the medial region, the perforating branches of the internal mammary arteries may even frequently encroach upon the origins of the sternocleidomastoid muscle and the anteromedial portions of the skin of the neck. It was mentioned previously that cutaneous branches arising from the anterior intercostal arteries are also found in this area.

These perforating branches exhibit peculiarities in their course through the subcutaneous tissue and skin that are at variance with the cutaneous branches of the intercostal arteries. In the latter, an arrangement corresponding to the thoracic vertebral segments is the rule; here, the same is found only by way of suggestion, and certain individual arteries supply large surfaces of the territory, far beyond the boundaries of the region one would like to attribute to them

32. Hyrtl, Corrosion Anatomy 91, p. 187.
33. Henle, p. 159.
34. C. FR. TH. Krause, p. 627.
35. Henle, p. 129; Sappey, p. 617.

by analogy to the cutaneous branches of the intercostal arteries. This fact is mentioned in some anatomy handbooks. Nevertheless, the assertions about the very strongly developed branches do not appear to be entirely correct. According to Krause,[36] the branch arising from the third intercostal space is supposed to be the most important one. Henle[37] indicates the same. The investigations of the author did not confirm this assertion in a single case. It was always the artery associated with the second intercostal space that was the largest and nourished the widest territory.[38]

The trunk of this artery travels in a general lateral and horizontal direction. On reaching a point midway between its origin and the axilla, or shortly before this point, it turns inferiorly at a right angle and descends in a highly tortuous course to the nipple. In its terminal portion, it surrounds the lateral[39] and occasionally the medial hemicircumference of the areola in an arc-like shape. Several twigs arise from the concavity of this arc toward the center of the nipple in radial fashion and terminate there in fine arterioles. Throughout its course this artery gives rise to numerous branches. Immediately at its exit point from the pectoralis major muscle it gives rise to numerous small branches that, medially oriented, supply the corresponding area of skin of the anterior surface of the sternum. They anastomose with the corresponding vessels from the opposite side. Other small cutaneous branches (one to three) travel short distances superiorly, inferiorly, and laterally, supplying small territories of the medial border areas of the skin overlying the pectoralis major muscle.

Like the sternal branches described above, these branches often leave the main trunk while still within the substance of the muscle and then appear as small independent cutaneous branches when viewed casually.

In its more distal horizontal course the main trunk gives rise, superiorly and inferiorly, to one or more branches that vary in importance.

They anastomose with the perforating branches (from the internal mammary artery) of the first and third intercostal spaces, which are highly variable in their degree of dominance. The perforating artery of the first intercostal space, when fully developed, encroaches for a considerable distance on the clavicular origin of the pectoralis major muscle.[40] When, as is frequently the case, it is only a small branch, the superior ramifications of the perforating branch associated with the second intercostal space take over the supply of this region. In most cases it is a single branch that has a larger ramification. The same holds true for those inferiorly directed branches of the main horizontal trunk. If the perforating branches corresponding to the third and fourth intercostal spaces are very small, then those branches of the horizontal trunk often gain a very important expansion. In one case a branch was even observed to advance down to the nipple and take part in its nourishment.

36. Krause, p. 607.
37. Henle, p. 129.
38. Compare Figures 3 and 4 (Plate II).
39. See Figures 3 and 4 (Plate II).
40. Figure 4 (Plate II).

At its point of downward angulation, the main trunk of this artery regularly gives rise to a strong branch that continues laterally, supplying an extended territory of skin through abundant arborizations. These form an anastomotic network with the cutaneous vessels from the thoracic branches of thoracoacromial artery and of the supreme thoracic artery.

A reciprocal relationship exists between these arteries. When the supreme thoracic artery is strongly developed, the territory it supplies is reduced, and vice versa. Many medial and lateral branches are given off by the descending portion of the artery, and they generally travel in a horizontal direction. The medial branches anastomose with the cutaneous vessels of the perforating branches from the internal mammary and intercostal arteries. The lateral branches, together with those branches from the thoracic arteries, form a rich anastomotic network.

One thoracic artery,[41] which is not mentioned at all in the anatomical literature, is particularly involved in the formation of this network. This artery was found in almost all cases investigated by the author. Moving away from the axilla, it descends on the lateral edge of the pectoralis major muscle to the fifth or sixth rib. One could aptly call it the superficial thoracic artery. It has in common with the other thoracic arteries that it may be of extremely variable origin. In many cases it arises directly from the axillary artery. It can sometimes be the most lateral or the most medial thoracic artery, or it may originate in the middle of the others. It may emerge together with one or more thoracic arteries from a common trunk, it may arise from the inferior subscapular artery, or it may originate directly from the axillary artery at its distal end (compare Figure 4, Plate II). Depending on this origin, the point at which it courses around the lateral edge of the pectoralis major muscle and appears on the anterior surface may naturally vary as well. It gives rise to lateral and medial branches. The lateral ones, especially numerous in the inferior portion of the territory, course to the lateral thoracic wall. The medial ones are much more numerous and larger. They course toward the median line, parallel to the ribs, until they anastomose with the terminal vessels of the perforating branches from the internal mammary arteries and with the intercostal arteries (see above). As a rule one to three branches go directly to the mammary tissue (Figures 3 and 4, Plate II).

The long thoracic artery also is important in supplying the skin of the anterior breast. Individual branches pass through the pectoralis major muscle and terminate in the deeper layers of the subcutaneous fat. The superolateral corner of this region is supplied by the supreme thoracic artery and the thoracic branch of the thoracoacromial artery.

The participation of the supreme thoracic artery, which Henle, Arnold, and others mention, is variable and probably never of any great importance. The thoracic branch of the thoracoacromial artery supplies the superior corner of this region more consistently with numerous branches that course toward

41. Compare Figure 2 (Plate I), Figures 3 and 4 (Plate II).

the median line. It has been mentioned previously that in its ramifications it anastomoses with the perforating branches of the internal mammary artery of the second intercostal space. In some cases this cutaneous branch emerges from the deltopectoral groove.

The anterior abdominal wall[42] is closely linked with the arterial supply of the breast. The entire territory of the anterior abdominal wall above the umbilicus, bounded laterally by the costal arch, shares with the breast the same main sources of arterial supply to the skin, namely the internal mammary artery. In the lower region, bounded by the inguinal ligament and the iliac crest, the vascularity of the femoral artery predominates. There are some notes in the anatomic literature about the cutaneous supply of this inferior portion, but they are fragmentary and incomplete, as will be brought out in more detail in the course of this presentation. In the literature there is practically no mention of the cutaneous arteries of the superior region.

Therefore, special attention was given to this area in the author's investigations.

In the majority of cases the following results were obtained. In the area of the sixth intercostal space the internal mammary artery divides into its terminal ramifications. At this point it usually gives rise to its last perforating branch. This branch either emerges directly on the surface at the medial edge of the insertion of the pectoralis major muscle, or it may penetrate the costal origin of the rectus abdominus muscle and the anterior rectus fascia. Shortly thereafter, it divides into two branches.

The smaller of the two branches courses laterally in a horizontal or slightly descending direction. Its course generally corresponds to the boundary of the costal origins of the pectoralis major and rectus abdominus muscles. Often it encroaches more onto the pectoralis major. With its ramifications, it supplies a narrow border territory of the breast and abdominal skin that corresponds to the sixth intercostal space, often up to the vicinity of the slips of the serratus anterior muscle. There its terminal ramifications anastomose with the anteriorly oriented cutaneous vessels from the lateral perforating branches of the intercostal arteries, which were described above.

The other, larger branch descends on the rectus sheath and divides again into two equal branches that course along the two borders of the sheath.[43] They give rise to a number of horizontally or diagonally descending lateral and medial branches that anastomose with the anterior cutaneous branches from the intercostal arteries.

These descending branches course to the region of the umbilicus with their terminal ramifications; there they anastomose with the branches of the

42. Compare Figure 2 (Plate I), Figure 4 (Plate II), and Figure 5 (Plate III).

43. The above-mentioned lateralward coursing main branch of our artery sometimes arises from the lateral one of these two descending branches as a side branch (compare Figure 4, Plate II).

superficial epigastric artery that ascend from below. The former shows a great similarity to the latter vessel. As the superficial epigastric artery is the principal supply of the skin of the lower abdominal region to the level of the navel, so the descending branches are the main source of nutrition for the skin of the upper abdominal region. The descending branch shares a relationship to the superior epigastric artery similar to that of the superficial epigastric artery with the inferior epigastric artery. Thus it follows that this previously undescribed artery can be called the superficial superior epigastric artery, and that what was up to now called the superficial epigastric artery can be called the superficial inferior epigastric artery instead.

In addition to the superficial superior epigastric, the superior epigastric artery participates in the cutaneous supply of this region.

Four, five, or even more cutaneous branches from this artery perforate the anterior wall of the rectus sheath.[44] In the majority of cases they are unimportant and pass into the anastomotic network of the superficial epigastric artery after a short distance without showing any characteristic arrangements or directional relationships. These arteries exist in reciprocal relationship with the medial descending branch of the superficial superior epigastric artery. If this branch is poorly developed, it is replaced by one or more larger branches arising from the superior epigastric artery, which may then correspond closely in their course and relationships to what was described for the descending branches. The participation of the cutaneous branches from the intercostal arteries on the lateral boundaries of the region has been discussed previously.[45]

The skin of the lower abdominal region is primarily supplied by the superficial inferior epigastric artery (superficial epigastric artery, subcutaneous abdominal artery of the other authors).[46]

It arises from the anterior surface of the femoral artery, emerges through the crescentic notch or through the upper portion of the aponeurosis bordering this notch,[47] and ascends to the superficial fascia of the abdomen. It spreads out over the fascia within the subcutaneous tissue.[48]

A few centimeters beyond the inguinal ligament, it divides into two main branches.[49] One branch ascends vertically with many convolutions as far as the navel, where it gives off its terminal ramifications. The other courses superolaterally in a diagonal direction and spreads out into the skin of the lateral abdominal wall. In the investigations of the author, a different course for this lateral branch was found several times. It coursed parallel to the lateral edge of the rectus abdominus muscle up to height of the navel while giving rise to

44. Figure 4 (Plate II): es.
45. The sixth (seventh) to the eleventh intercostal arteries have connections to the skin of the abdomen.
46. See authors, especially Sappey, pp. 58–59.
47. Henle, p. 190.
48. According to Henle (p. 190) it spreads out into the superficial fascia itself.
49. Theile differentiates an abdominal branch and an iliac branch. Theile, p. 223.

numerous diagonally ascending branches in medial and lateral directions. The lateral branches were particularly well developed and spread to the lateral abdominal wall. However, none was sufficiently strong to be designated the main branch, in contrast with the vertically ascending branch [see the figure below].

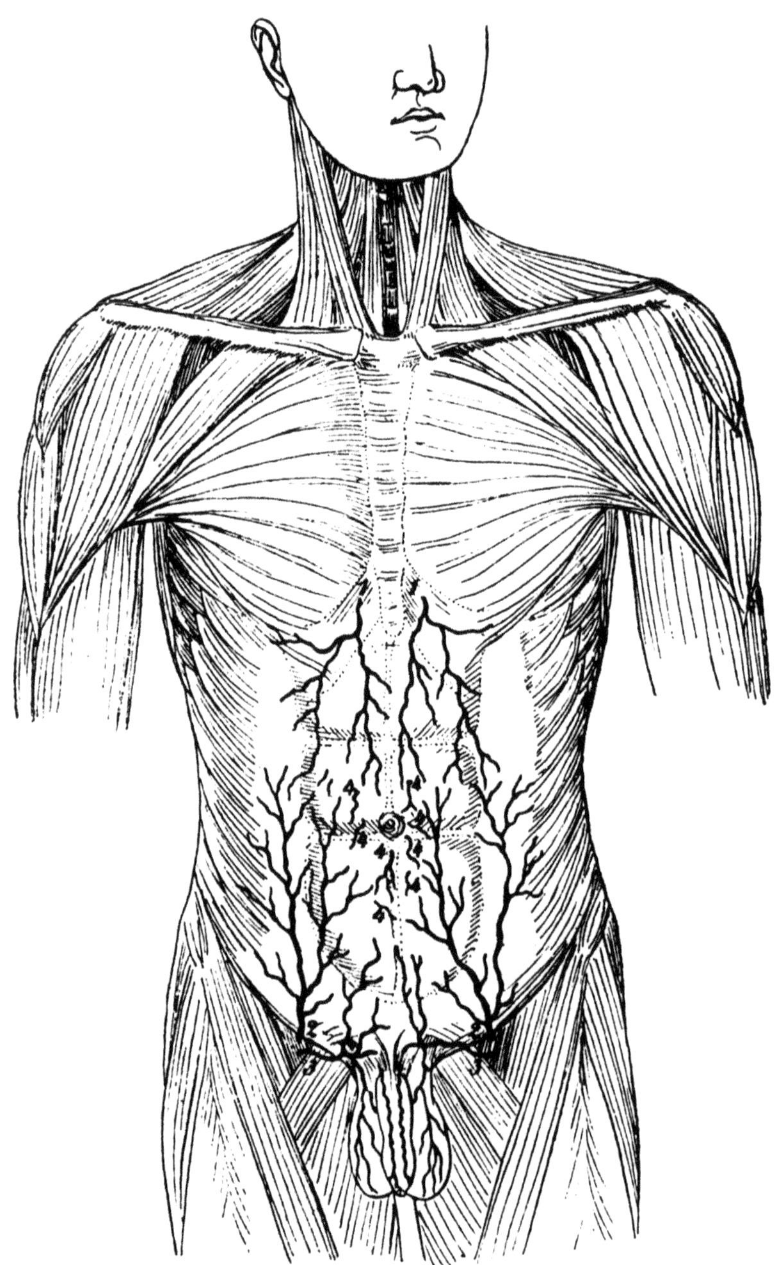

1 Superficial superior epigastric artery.
2 Superficial inferior epigastric artery.
3 External pudendal artery.
4 Perforating branches of the superior and inferior epigastric arteries.

The first branches of the superficial inferior epigastric artery are destined for the lymphatic glands and the skin of the inguinal region. Sappey divides the remaining cutaneous branches into three groups: (1) medial branches, which anastomose with cutaneous branches from the inferior epigastric artery; (2) lateral branches, which anastomose with branches from the circumflex iliac artery; and (3) central branches, which become confluent with arterial branches from the lumbar arteries.

It has already been noted that the superficial superior epigastric artery, not mentioned by Sappey, is also strongly involved in the formation of this anastomotic network.

Between the two medially ascending branches of the superficial inferior epigastric artery, a number of smaller cutaneous branches perforate the superficial fascia of the abdomen at exit points that are not particularly characterized. They converge upon the umbilicus and anastomose with the network of arteries surrounding the umbilicus. These small arteries originate from the inferior epigastric artery.[50]

Sappey differentiates between lateral and medial branches according to their direction of departure from the vertically ascending terminal portion of this artery. The lateral ones perforate the superficial fascia of the abdomen at the lateral edge of the rectus abdominus; the medial ones pass in greater number than the former into the skin at the medial edge of this muscle and terminate in the area of skin in front of the linea alba and the medial edge of this muscle.

Their exit points extend downward no further than a horizontal line passing through the midpoint between the umbilicus and the root of the penis.[51]

The lateral abdominal region is supplied by the lateral perforating branches of the lumbar arteries as far as the territory of the dorsal branches.[52] The uppermost lumbar artery follows the lower edge of the twelfth rib,[53] while the fourth travels along the iliac crest. The anterior branch (abdominal branch) of these arteries passes anteriorly either in front of or behind[54] the quadratus lumborum muscle. On reaching the lateral edge of this muscle, it divides into two branches, deep and superficial, both of which advance toward the terminal ramifications of the inferior epigastric artery. The superficial branch courses within the space between the internal and external oblique muscles. From it emerge a variable number of cutaneous arteries that make their way through the external oblique muscle. On reaching the surface of the abdominal fascia, they travel toward the median line in a horizontal or slightly downward direction, to merge with the anastomotic network of the superficial inferior epigastric artery. They function very similarly to the cutaneous branches from the intercostal

---

50. See Figure 4 (Plate II): ei.
51. Author's observation.
52. Sappey, p. 666.
53. Sappey, p. 568.
54. Henle states that, as a rule, the upper lumbar arteries lie deep to the muscle, while the lower ones lie superficial to it. Henle, p. 163.

arteries in terms of their relationship to the corresponding vertebral segments, and with the same modification and variations. It is sufficient to refer to what has already been stated concerning the cutaneous branches of the intercostal arteries.[55]

The inferolateral abdominal border region is supplied neither by the branches of the superficial inferior epigastric artery nor by the lumbar arteries. The superficial circumflex iliac artery is destined for this territory. This artery originates from the femoral artery, approximately at the same height as the superficial inferior epigastric artery. Often it is a branch of the latter[56] (see Figure 4, Plate II; Figure 5, Plate III). It travels laterally, parallel to the inguinal ligament as far as the anterior superior iliac spine, enclosed by the abdominal fascia. Numerous branches of the superficial circumflex iliac artery, traveling at an acute angle parallel to the superficial inferior epigastric artery or oriented diagonally lateralward, supply the cutaneous territory lateral to the region of the superficial inferior epigastric artery. Descending branches pass to the skin of the upper part of the thigh. On reaching the anterior superior iliac spine, the artery divides into numerous diverging terminal branches that anastomose with cutaneous branches from the gluteal artery, the circumflex femoral artery, and the fourth lumbar artery.

In the above discussion, the cutaneous arteries of the entire abdominal region have been described with the exception of one area in the lower medial region. It is represented by a quadrangle of unequal sides.[57] The longer diagonal of the figure extends from the middle of the linea alba to the root of the penis (suspensory ligament). The shorter diagonal is defined by connecting two points located 1–2 cm lateral to the external inguinal ring on the inguinal ligament on the two sides. There is no information in the anatomic literature about the blood supply of the major portion of this area, i.e., that cutaneous territory that does not belong directly to the genitalia. In the investigations of the author, it became apparent that this territory is supplied exclusively by previously unde-scribed abdominal branches of the external pudendal arteries.

Of the two arteries, both of which one generally finds bilaterally,[58] it is the more superficial one, lying closer to the inguinal ligament, that almost exclusively supplies the cutaneous arteries to this region (the author). Shortly before it reaches the inguinal ligament and about 2 cm from the external inguinal ring, it gives rise to two abdominal branches, an inferior one and a superior one, which are of some significance.[59] The inferior one continues in a horizontal direction along the mons pubis and passes over the spermatic cord while giving rise to cutaneous branches at the corresponding area of the scrotal skin. On either side of the spermatic cord, it breaks up into several diverging terminal

55. See Figure 2 (Plate I), Figure 4 (Plate II).
56. Krause, p. 653; Henle, p. 190.
57. The author.
58. Henle, p. 190.
59. Figure 5 (Plate III).

branches. They supply the territory of skin corresponding to the symphysis as far as the root of the penis (pubic ramus of the superior external pudendal artery). The superior branch ascends diagonally toward the center of the linea alba. Along this path it gives rise to numerous medial and lateral branches, all of which ascend diagonally. The medial ones course toward the linea alba, their terminal ramifications anastomosing with the corresponding vessels on the opposite side. The lateral branches are anastomotically connected with medial branches of the superficial inferior epigastric artery. In some cases, this abdominal branch of the superior external pudendal artery, as one could call it, is subject to some variation. It is rare that it does not completely fill the specified area. Nevertheless, lateral branches often extend far beyond the borders into the territory of the superficial inferior epigastric artery.

This final territory of abdominal skin is thus intimately related to those of the genital and perineal areas. Therefore the description of this territory carries over directly to this final portion of the cutaneous arteries of the trunk. The difference between the sexes requires a separate description.

The cutaneous area of the male inguinal and genital region[60] is divided for practical reasons into an anterior (ventral) and a posterior (dorsal) region. The more anterior encompasses the skin of the penis and the anterior portion of the scrotal skin; the more posterior encompasses the posterior portions of the scrotal skin as well as the perineal and perianal skin.

The anterior region is supplied for the most part by the external pudendal arteries. As already mentioned, one generally finds two of these arteries, namely the superior external pudendal artery and the inferior external pudendal artery, which may occasionally arise from a short common trunk.[61]

The former arises from the medial side of the femoral artery about 2–4 cm inferior to the inguinal ligament. It travels medially and perforates to the surface through an opening in the pectineal fascia after traveling a short distance. In the subcutaneous tissue between the skin and fascia it approaches the inguinal ligament. Along this course it gives rise to ascending and descending branches. The former is spread out in the triangular area of skin between the inguinal ligament and the root of the artery. The descending ones anastomose with cutaneous branches of the inferior external pudendal artery. Shortly before it reaches the inguinal ligament, it gives rise to the two abdominal branches as noted above. The descending branch soon resolves into terminal branches that participate in supplying a territory of skin involving the penis and the anterior portions of the scrotum.

As a rule, the inferior external pudendal artery originates from the femoral artery somewhat more inferiorly than the superior external pudendal artery.[62] Its proximal portion is located anterior to the femoral vein and behind the

---

60. Compare Figure 5 (Plate III) and Figure 6 (Plate IV).
61. Sappey, p. 671; Henle, p. 190.
62. Sometimes it arises from the profunda femoris. Sappey, p. 671.

saphenous vein shortly before these two converge. Between the pectineal muscle and the fascia lata it travels medially in a horizontal direction. After it has reached the adductor longus muscle, it passes through the fascia and enters the subcutaneous tissue of the upper part of the thigh in its passage toward the testicle.

Along this pathway it gives rise to numerous secondary ascending and descending branches, of which the ascending ones have already been described. The descending ones, four to five in number, supply a portion of the skin of the upper thigh with abundant ramifications; the lower boundary of this thigh territory is defined by a line passing from the point of emergence of the femoral artery below the sartorius muscle, parallel to the inguinal ligament, and up to the gracilis muscle (see Figure 5, Plate III). After the main trunk of the inferior external pudendal artery has reached the scrotal skin, it divides near that point into its terminal ramifications.

In the author's investigation, the arrangement of the cutaneous arteries of the penis and of the anterior region of the scrotum proved to be somewhat different than is usually described. According to the information from anatomy textbooks,[63] it appears as if both external pudendal arteries supply this territory in the same fashion. In the cases investigated by the author, it more often was the case that the superior external pudendal artery supplied a restricted proximal portion of the scrotal skin in addition to the previously described medial abdominal region. The entire remaining portion of the area fell to the inferior external pudendal artery, while the superior one took part only indirectly by means of anastomosis. The somewhat complicated relationships of these cutaneous arteries may perhaps best be illustrated through the description of a particular case. Figure 5 (Plate III) depicts the anterior projection of an especially successful injection of this territory of skin. The superior external pudendal artery has nearly terminated in giving rise to abdominal branches, which were particularly strong in this case. It ends with a cutaneous branch of average strength that participates in supplying the anterior scrotal wall.

The inferior external pudendal artery reaches the scrotum 1–2 cm inferior to the superior external pudendal artery. In the loose connective tissue between the scrotum and the spermatic cord, it approaches the root of the penis. Along this course it gives rise to three to five descending branches. These anterior scrotal branches[64] move in a vertically descending direction to the base of the scrotum. They divide in two at acute angles into terminal branches that ultimately anastomose at the base of the testicle with corresponding cutaneous branches arising from the internal pudendal artery. On the lateral walls of the scrotum they anastomose with cutaneous branches from the obturator artery.[65] That is,

<hr>

63. For example, Henle, p. 191; Sappey, p. 671.

64. Theile, p. 224.

65. Author's observation. Krause does not mention this in his tabular overview of the blood vessels of the individual organs. Theile mentions cutaneous branches from the obturator artery at the scrotum and labia majora, respectively, but without greater detail. Theile, p. 203.

the obturator artery, after leaving the obturator canal, divides into a posterior and an anterior branch. The anterior branch,[66] the only one of interest here, passes between the adductors, the strongest branch passing between the adductor brevis and longus muscles. It supplies the pelvic origins of these adductors and also the pectineal and gracilis muscles. It gives rise to cutaneous branches, especially in the region of the gracilis muscle, as well as in the uppermost region of the medial thigh surface (medial to the cutaneous branches coming from the inferior external pudendal artery), and also in the lateral portions of the scrotal skin. These lateral scrotal branches, as they can be called, are distributed analogously to the branches of the inferior external pudendal artery described above.

After giving rise to the anterior scrotal branches, the inferior external pudendal artery reaches the surface of the penis. It then divides into several (two to four, three as a rule) branches, which descend subcutaneously in the penis along with the dorsal artery of the penis parallel to the prepuce, supplying the lateral and posterior (associated with the testicle) portions of the penile skin. Through numerous anastomoses they are interconnected with each other and with the dorsal artery of the penis. They descend until they reach the arterial arcade (formed by the two dorsal arteries of the penis) that passes into the circular groove of the glans penis.

The dorsal artery of the penis arises as one branch (with the deep penile artery) coming from the penile artery shortly after its emergence through the urogenital diaphragm. Immediately adjacent to the medial suspensory ligament of the penis it reaches the dorsal surface of the penis; here it gives rise to branches that pass to the skin of the root of the penis and the upper portions of the scrotum,[67] and it then courses in the medial groove toward the glans penis; the dorsal arteries of the penis are separated from one another by the dorsal vein but are in anastomotic contact through numerous horizontally oriented fine twigs (the author).[68] Shortly before they reach the glans, they diverge and pass into the circular groove that separates the glans from the rest of the penis and then pass laterally around the penis. On the ventral surface of the glans they pass into the interior of the glans and flow together to form an arterial arcade.

In the superficial portion of this arcade, as has already been described, the branches arising from the external pudendal artery anastomose with their terminal ramifications. Numerous fine cutaneous vessels passing to the glans and prepuce[69] arise from this arcade.

One of the branches into which the inferior external pudendal artery divides surrounds the root of the penis in an arc and advances to its ventral surface. Shortly before reaching the median line, it divides in two, one branch

66. Henle, p. 178; Sappey, p. 687.
67. Theile, p. 209.
68. See Figure 5 (Plate III).
69. Arnold, p. 533.

oriented toward the tip of the penis and the other passing toward the root of the penis and scrotum, and functioning just like an anterior scrotal branch.

Naturally, all cutaneous branches of the penis exhibit a strongly tortuous course under nonerectile conditions.

The posterior scrotal arteries, as already noted, are a part of this network.[70] They are associated with the territory of the internal pudendal artery, the dorsal penile arteries of which have already been described. The internal pudendal artery[71] becomes associated with the skin of the perineum and genitalia beginning at that point where it enters the ischiorectal fossa between the sacrotuberous ligament and sacrospinous ligament.[72] It passes around the lateral edge of the ischiorectal fossa on the medial surface of the ischium and is covered by the falciform process of the sacrotuberous ligament in its course. On reaching the lateral insertion of the superficial transverse perineal muscle, it divides at an acute angle into its two terminal branches, the penile artery and perineal artery. Along its course through the ischiorectal fossa the internal pudendal artery gives rise to lateral and medial branches. The lateral branches reach the muscles originating at the ischium, especially to the inferomedial edge of the gluteus maximus muscle and to a rather extensive cutaneous territory of this region; these lateral branches will be discussed below.

The medial branches, also known as the external hemorrhoidal arteries, leave the common pudendal artery with two to four small stems (the author).[73] Following a tortuous course, they pass with many branches through the adipose tissue above the levator ani muscles and form a rich arterial network supplying the levator ani muscle and the external anal sphincter as well as the skin covering this area, along with numerous anastomoses with branches from the opposite side.

According to Theile, the inferior gluteal artery also participates in supplying the skin of this region. According to him, a more important cutaneous branch of the inferior gluteal artery usually travels anterior to the sacrotuberous ligament diagonally in the direction of the coccyx, spreading out in the adipose tissue of the perineal fossa (anastomosing with branches of the internal pudendal artery).

Individual laterally directed branches also emerge from the external hemorrhoidal arteries,[74] which reach the inferomedial edge of the gluteus maximus muscle and arborize in the skin of this region.

The territory of the superficial perineal artery[75] adjoins anteriorly the cutaneous territory of the external hemorrhoidal arteries. Running parallel to

<hr>

70. Theile, p. 208.

71. Common pudendal artery (Gegenbaur).

72. Compare Henle, p. 182+; Sappey, pp. 660–663.

73. According to Henle (p. 182), with one to three small branches. According to Sappey (p. 660), with three to four small branches. Theile, p. 208.

74. Author's observation. See Figure 6 (Plate IV).

75. Perineal artery (Henle). Transverse perineal artery.

the superficial transverse perineal muscle, this artery courses toward the median line and divides into its numerous terminal branches, some of which enter the dartos fascia and some of which pass deep to it to the bulbocavernosis muscle. Along its course parallel to the superficial transverse perineal muscle, the superficial perineal artery gives rise to medial and lateral branches.[76] The medial branches vary in number (two to three as a rule) and course into the arterial network that supplies the external anal sphincter and the perianal skin. They comprise the anterior portion of this arterial network and anastomose with the corresponding branches from the opposite side.

The lateral branches course to the ischiocavernosis muscle[77] and a rather extensive cutaneous territory on the posteromedial surface of the upper thigh (author). They anastomose with cutaneous branches from the internal pudendal artery, the external hemorrhoidal arteries, and the obturator artery (*see above*).

The terminal ramifications of the perineal artery, the posterior scrotal arteries (three to four in number), descend in parallel, highly convoluted courses to the posterior surface of the scrotum.

They are connected with one another through an anastomotic network. They divide at acute angles into numerous terminal branches that merge with the arterial network at the base of the scrotum, as previously noted.

They thus come into contact with the previously described scrotal branches coming from the inferior external pudendal artery and the obturator artery. Branches of the posterior scrotal arteries also penetrate the posterior portion of the scrotal septum.[78] Theile describes yet another branch of the medial circumflex femoral artery for this area, a branch that allegedly supplies the adductor muscle origins, to the external obturator muscle, and to the superoposterior portion of the scrotum, anastomosing there with branches coming from the internal pudendal and obturator arteries.

In the female,[79] the arrangement of the cutaneous arteries in the genital and perineal regions deviates only slightly from what has been said thus far, corresponding to the differences in the external genitalia.

On close examination it becomes apparent that despite some modification in detail, the same basic arrangement may be recognized. One has only to compare those structures that have homologous developmental histories.

In the female, the nutrition of the skin by the internal pudendal artery, insofar as it concerns the ischiorectal fossa, corresponds exactly to what was described for the male anal region, the same external hemorrhoidal arteries and

---

76. Sappey, p. 661.

77. See Figure 6 (Plate IV).

78. Septal scrotal and marginal scrotal arteries (Barkow). M. J. Webber mentions minor posterior scrotal arteries (compare Henle, p. 183), which come out of the transverse perineal artery and the major posterior scrotal artery which, in turn, arises from the stem of the internal pudendal artery. The author of this work did not succeed in finding the major posterior scrotal arteries in a single case.

79. Compare Figure 7 (Plate IV).

their lateral branches spreading into the cutaneous territory of the upper thigh.

The perineal artery, which in the male supplies the posterior wall of the testicle and the bulbous urethra, supplies the corresponding structure in the female[80] with its branches. The posterior labial arteries (corresponding to the posterior scrotal arteries) run to the posterior portion of the labia majora, to the small labia of the vulva, and to the posterior portion of the vestibule. "The clitoral artery of the female, which runs to the clitoris, corresponds to the penile artery in its course and branchings and differs from it only in the uniformly smaller caliber of all of its branches."[81] The cutaneous branches from the external pudendal arteries also show a great similarity to those of the male. The superior external pudendal artery gives rise to branches at the clitoral prepuce in addition to abdominal branches. Breaking up into numerous terminal branches, the inferior external pudendal artery supplies the anterior portion of the large labium of the vulva (anterior labial arteries) and forms, along with the branches of the posterior labial arteries, an arterial network similar to that found in the male at the base of the scrotum. Here also, cutaneous branches coming from the obturator artery contribute to this network.

80. Henle, p. 184; Sappey, p. 663; Arnold, p. 531.
81. Henle, p. 185.

Chapter II

Cutaneous Arteries
of the Neck

(Compare Figure 8, Plate V)

The description of the cutaneous arteries of the neck is related to the description of those of the trunk, since the neck is actually only a modified portion of the trunk. A description of the skin of the nuchal region is excluded here because it fits in more easily in the discussion of the cutaneous arteries of the upper extremity, to which the major portion of the musculature of the nuchal region belongs.

The neck region to be considered here is bordered: (1) at the chest by the clavicle; (2) at the head by the inferior edge of the mandible, by the parotid gland, and by the cranial insertion of the sternocleidomastoid muscle; and (3) at the lateral boundaries by the lateral edges of the trapezius muscles.

This region again divides into an inferior and a superior territory, the former associated with the thyrocervical trunk and the transverse cervical artery, and the latter associated with the external carotid artery. The border between the two extends laterally from the middle of the thyroid gland parallel to the clavicle to the posterior (dorsal) edge of the sternocleidomastoid muscle. Then it angles superiorly following this edge of the sternocleidomastoid. Arriving at the level of the border of the mandible, it angles posteriorly to the median line in the nuchal area.

The cutaneous arteries of the inferior region, normally fairly constant in their course and direction, are subject to great variation in their origins. One can scarcely find two cases where the origins of these cutaneous arteries are identical. The basis of this phenomenon is the great inconsistency of the secondary branches of the subclavian artery from which these cutaneous branches originate. "Often all of these branches crowd together into one narrow space between the origins of the subclavian artery and the medial edge of the anterior scalene muscle, whereas in other cases the section of the artery lying behind the anterior scalene muscle gives rise to a variable number of branches, and individual branches may even extend out over the lateral edge of the anterior scalene muscle."[82]

To achieve some measure of orientation in this confusing array of individual variations, it is of value to establish some sort of arrangement as a norm and then to investigate the arrangement of the cutaneous arteries within this norm.

On the basis of his abundant experience, Henle establishes the following arrangement as the standard:

> On the medial edge of the anterior scalene muscle but arising from the anterior wall of the artery, four branches take origin in the middle of a common stem, the thyrocervical trunk, and spread out in the neck in three opposite directions— medially, superiorly, and laterally. The inferior thyroid artery passes medially, the ascending cervical artery ascends superiorly to the deep anterior neck muscles, and the lateral component is shared by two vessels, the superficial cervical and transverse scapular arteries, the former to the edge of the trapezius muscle and the latter to the lateral corner of the scapula. The transverse cervical artery originates on the other side of the anterior scalene muscle.[83]

For this norm, which is actually a very frequently occurring combination, the cutaneous arteries behave approximately as follows:[84]

The inferior territory of this inferior neck region is supplied by the transverse scapular artery. At the point where this artery enters the supraclavicular fossa from beneath the lateral edge of the sternocleidomastoid muscle, it gives rise to several cutaneous branches that enter the lower portion of the platysma and the skin of that region. As a rule these cutaneous branches show the following arrangement. A short stem branches[85] from the inferior side of the transverse scapular artery at that location and descends to the middle of the superior edge of the clavicle. One could call it the subcutaneous supraclavicular artery. Shortly before it reaches the edge of the clavicle, it divides into two (or three) diverging branches as an anterior and posterior branch. The anterior branch turns medially and runs parallel to the upper edge of the clavicle about as far as the sternoclavicular joint. It sends off descending and ascending branches; the descending ones, three to five in number, descend over the clavicle to the anterior chest wall, in part enclosed in the platysma muscle and in part superficial to it. They supply the upper peripheral region of the anterior pleura, sometimes

---

82. Henle, p. 119.

83. Henle, p. 122.

84. The cutaneous arteries of the anterior and lateral pharyngeal region are hardly mentioned in the anatomical literature. The following presentation is based on observations of the author.

85. Figure 8 (Plate V).

to the entire extent of the first intercostal space. The ascending branches (two to three) turn toward the cranium parallel to the direction of the fibers of the sternocleidomastoid muscle and reach, when very strongly developed, the upper boundary of the region.[86] Worthy of note is the most medial of these branches, which nourishes the inferomedial portion of the anterior neck with its terminal branches. It is either very strongly developed or is completely lacking. In the latter case it is replaced by the cutaneous branch of the first perforating branch of the internal mammary artery,[87] which emerges between the clavicle and the first rib. It ascends over the sternoclavicular joint and arborizes in the skin overlying the sternal portion of the sternocleidomastoid muscle.[88]

The posterior branch of the subcutaneous supraclavicular artery from the transverse scapular artery courses laterally as far as the lateral edge of the trapezius muscle or even to the vicinity of the acromion. Its numerous descending branches spread to the clavicular attachments of the pectoralis major and deltoid muscles. They anastomose with cutaneous branches of the thoracoacromial artery. Individual ascending branches disappear in the subcutaneous tissue and the skin of the supraclavicular fossa.

Sometimes, as has already been mentioned, a third diverging branch goes out from the cutaneous trunk of the transverse scapular artery. It is very inconstant and never of any great importance. From the point of division onward it courses superiorly and spreads out partially within and partially deep to the platysma as far as the upper boundary of this lower neck region. Only inconsequential small branches emerge through the platysma and run to the skin.

In its further course through the supraclavicular fossa the transverse scapular artery gives rise to still more small cutaneous branches. These penetrate the superficial cervical fascia and disappear in the fatty tissue, the lymphatic glands, the platysma muscle, and the skin of this territory through ascending pathways. They are not very constant and are often more or less replaced by similar branches coming from the transverse cervical artery.

This is the only territory where the transverse cervical artery participates in the arterial supply of the lateral neck region in that arrangement described as normal by Henle. When, on the other hand, it exchanges its point of origin

86. Compare Figure 8 (Plate V).
87. Compare Figure 3 (Plate II).
88. Arnold, p. 485; Henle, p. 130. Also compare Theile, p. 124. A very similar relationship has been described by Arnold and Henle for the muscular branches of these arteries. One branch from the first perforating artery supplies the sternal origins of the anterior pharyngeal muscles with its branches. The fact that the cutaneous artery described by the author is not merely a branch of this muscular branch is explained by the following: The muscular branch does not emerge onto the surface of the anterior thorax between the clavicle and the first rib; it ascends behind the clavicle and arrives at a more superficial position superior to it between the sternal and clavicular origins of the sternocleidomastoid muscle, while the cutaneous branch passes over the sternoclavicular joint to the anterior neck, lying on the platysma. Naturally, both are subsidiary branches of the same artery (first perforating branch).

with that of the transverse scapular artery,[89] as often happens, then the cutaneous arteries described for it are correspondingly replaced, with slight modifications.

In the central area of the lower lateral neck region, two or three cutaneous arteries are especially noteworthy because of their unique arrangement. Inferior to the lateral edge of the sternocleidomastoid muscle, they usually travel laterally deep to the platysma as far as its lateral edge; they then wrap themselves around the lateral edge, reverse to a medial direction, and continue just superficial to the platysma parallel to the clavicle. They then ramify in the as yet unsupplied anterior portion of the lower neck region. As consistent as these cutaneous arteries are in their course and orientation, their origins differ. All of the variations in pattern and origin are also expressed in this region. In the case set forth as the norm, the more inferior of these two arteries arises from the transverse cervical artery and the more superior from the superficial cervical artery.[90]

The remainder of the lower neck skin is part of the territory nourished by the superficial cervical artery.

This is the somewhat irregular quadrilateral region lying between the lateral edges of the trapezius muscle, the platysma, and the upper boundary line of the lower neck region.

About 2.5–3 cm above the clavicle, the superficial cervical artery[91] courses in a diagonal direction superficially and toward the back, first lying between the sternocleidomastoid muscle and the anterior scalene, then more superficially in the supraclavicular fossa deep to the platysma, and then as far as the anterior edge of the trapezius muscle beneath which it finally disappears. Along this course it gives rise to ascending and descending cutaneous arteries. The ascending ones, one to three in number, ascend with abundant ramifications into the territory of skin just described, supplying this territory and encroaching even into the skin overlying the trapezius muscle. These vessels supply the fat, the numerous lymphatic glands of the region, and the skin.

The descending cutaneous branches of the superficial cervical artery anastomose with the previously described ascending cutaneous branches of the supraclavicular fossa, which originate from the transverse scapular and transverse cervical arteries.

The upper neck region is supplied exclusively by the external carotid artery. Indeed, it is principally those arteries associated with the head region that nourish this territory. Only that narrow median region between the diverging medial edges of the platysma is supplied by arteries of the larynx and of the thyroid gland.

Of the arteries of the head, the submental artery possesses the most extensive territory. The submental artery[92] emerges from the facial artery at the point where the latter wraps around the inferior edge of the mandible and

---

89. Compare Henle, p. 122.
90. Figure 8 (Plate V): cs, tc.
91. Compare especially Arnold, p. 480; Henle, p. 133.
92. Henle, p. 96.

passes to the face. In the submandibular fossa, the submental artery passes toward the chin parallel to the anterior belly of the digastric muscle. Shortly beyond its origin it sends several branches to the submandibular gland. One of these,[93] larger than the others, penetrates the substance of the gland and appears on the surface of the platysma 1–2 cm inferior to the edge of the mandible.

Here it arborizes in the triangular cutaneous territory bounded anteriorly by the medial edge of the platysma, posteriorly by the anterior edge of the sternocleidomastoid muscle, and superiorly by the curve of the mandible. In one particularly clear dissection, it could be followed up to an anastomosis with the ascending cutaneous branches coming from the anterior cutaneous rami of the subcutaneous supraclavicular artery.[94] Smaller cutaneous branches[95] of the submental artery, varying in number and prominence, supply the skin of the lower chin region.

According to Theile[96] there are as a rule two to four branches that pass over the edge of the mandible into the face. The larger branches emerge from the platysma lateral to the insertion of the anterior belly of the digastric muscle, and the smaller branches emerge medial to it. Small cutaneous branches from the sublingual artery,[97] a branch of the lingual artery, are also regularly found in this region. These branches anastomose with the small cutaneous branches of the chin region coming from the submental artery; sometimes they even replace them entirely.[98] More rarely the reverse holds true, the branches from the sublingual artery being completely absent.[99]

More posteriorly, several cutaneous branches originating from the parotid rami of the external carotid artery interconnect with the branches arising from the submental artery.[100]

Ordinarily a larger branch penetrates the parotid gland about 1–2 cm inferior to the external auditory canal and then descends vertically crossing onto the superficial surface of the platysma muscle at the level of the mandibular angle and reaching the lower boundary of the upper region with its terminal branches. The superior portion of the region of skin located posterior to the parotid gland is associated with branches of the occipital artery. These arteries will be discussed in connection with the remaining cervical branches of the occipital artery.

The median region of the neck between the diverging edges of the platysma muscle is the territory of the laryngeal arteries. Branches of the superior thyroid artery constitute, together with the hyoid branch of the lingual artery,

93. Usually single, but occasionally two or three.
94. Compare Figure 8 (Plate V).
95. Arnold, p. 452; Sappey, p. 581.
96. Theile, p. 66.
97. Theile, p. 63.
98. Henle, p. 96; Theile, p. 63.
99. According to Haller, 9 times in 50 cases. Icones anatom. fasc. 3. p. 5. not. 9.
100. Compare Figure 8 (Plate V).

an extensive network[101] on the anterior surface of the thyroid cartilage, from which branches pass to the strap muscles. In addition, muscular branches emerge from the superior thyroid artery in variable numbers, sometimes arising directly from it and sometimes from the glandular branches. They supply the superior portions of the strap muscles. Individual cutaneous arteries penetrate these muscles at different exit points and ramify in the skin of this territory without exhibiting any uniform relationships in their organization and orientation.[102]

The author could find no trace of the cutaneous branches from the inferior thyroid artery that are mentioned by Krause.[103] Theile also does not mention any cutaneous branches arising from the inferior thyroid artery.[104]

101. For more detail see Henle, p. 93+.
102. The author. Also compare Theile, p. 59.
103. Krause, p. 274.
104. Compare Theile, p. 119.

Chapter III

Cutaneous Arteries
of the Head

(See Figures 8 and 9, Plate V)

The external carotid artery, whose relationships with the skin of the upper neck territory have already been described, supplies the entire skin of the head with the exception of a very small area associated with the internal carotid artery. At the point where it reaches the head region, the external carotid ramifies into a number of musculocutaneous branches. These branches are so prominent and so conspicuous that they have already been thoroughly described in some detail.

Reference to the anatomic literature is therefore sufficient. The complicated and interesting arrangement of the cutaneous arteries of the eyelids has also been set forth in exact detail through the work of Langer and Fuchs.[105]

A few new observations made by the author are noted in the footnote below.[106] Essentially there was left to the author's own research only the investiga-

105. Langer C. On the blood vessels in the eyelid. Wiener medic. Jahrbücher 1878. Fuchs E. On the anatomy of the blood and lymph vessels of the eyelids. Archiv fur Ophthalmologie 1878. Bd. 24.

106. New observations by the author on cutaneous arteries of the head: (a) Anastomoses of the chin cutaneous branches of the submental artery with smaller medial cutaneous branches of the external maxillary artery, which it gives off in the region from the point it winds around the inferior mandibular edge to where the inferior labial artery is given off. These branches, although fairly constant, are not mentioned anywhere (compare Figure 8, Plate V). (b) At the commissure, ascending cutaneous branches from the inferior labial artery often anastomose with descending cutaneous branches from the superior labial artery (compare Figure 8, Plate V). (c) The division of the transverse facial artery into its diverging terminal branches often takes place close to the emergence from the parotid gland. As a rule, one can distinguish three main branches: an upper one, a middle one and a lower one. (d) In place of the single zygomaticoorbital artery, which Tiedemann, Arnold, and Henle describe, the author found several (two to four) separate arteries in the majority of cases. Sappey also describes several such supraorbital branches, as he calls them.

tions of the cutaneous arteries of the external ear and the cutaneous arterial network of the scalp.

The anterior and posterior auricular arteries are described as the cutaneous vessels for the external ear. There is practically nothing to be found in the literature concerning a more detailed distribution of these arteries.

One can divide the skin of the pinna into two arterial territories, that of the anterior auricular artery and that of the posterior auricular artery. The former includes the superior and anterior portion of the pinna along with the tragus as far as the intertragal notch. The area of the posterior auricular artery encompasses the posterior helix and the lobule. On the concave and convex sides of the concha the territories blend through abundant anastomoses.

First, two or three anterior inferior auricular arteries pass from the superficial temporal artery to the anterosuperior portion of the external auditory canal, to the anteroinferior part of the helix, and to the tragus, arborizing there in diverging branches. Further, at the approximate level of the zygomatic process, two anterior superior auricular arteries originate from the temporalis muscle, occasionally from a short common stem. One of these passes over the anterior edge of the helix to the concave surface of the pinna where it ramifies into its terminal branches. One branch courses along the superior edge of the helix toward the back. The others supply the triangular fossa and then pass over the superior crus of the anthelix into the groove between helix and anthelix known as the scapha. The other anterior superior auricular artery courses over the elevation corresponding to the triangular fossa on the convex surface facing the skull and passes into the sulcus corresponding to the anthelix. It descends within this sulcus to anastomose with an ascending branch from the posterior auricular artery to form an arterial arcade at about the level of the helical spine. Throughout its entire course, it gives rise to numerous branches that supply the superior portion of the convex posterior surface and that anastomose with branches of the anterior artery around the posterior edge of the ear.

At first the posterior auricular artery, the principal artery of the second territory of the pinna, ascends behind the parotid gland and then passes vertically in the groove between the mastoid process and the conchal cartilage.

On reaching the lower edge of this groove, it gives off two branches for the external ear, often from a common stem.

The one anterior branch goes through a fissure formed by the helical tail and the antitragus or passes anterior to the latter, closer to the tip of the lobule, on the outer surface of the antitragus. There it gives rise to branches that pass toward the earlobe. Some of these travel over the intertragal notch and interconnect with the cutaneous arteries of the tragus from the superficial temporal artery. This main branch soon ramifies into an abundant terminal arborization. The majority of the terminal twigs ramify within the concha as far as the inferior edge of the anthelix (anastomosing with the auricular arteries from the superficial temporal artery.)

One or two branches ascend within the groove between the helix and anthelix into an arcade concave anteriorly, to flow with the auricular arteries of the superficial temporal artery at the entrance to the scapha.

The posterior branch of the posterior auricular artery emerges with two to three branches on the deep surface of the concha on its convex side.[107] In the groove corresponding to the anthelix the ramifications of these branches flow into an arterial arcade that continues superiorly into the arcade formed by the ramifications of the posterior auricular artery from the superficial temporal artery. Many branches pass out from this arcade around the posterior edge of the pinna, anastomosing with branches of the peripheral artery of the concave side running in the groove between helix and anthelix.

In addition to the cutaneous arteries of the external ear, the cutaneous arterial network of the scalp is worthy of examination because of the peculiarity of its organization and the lack of prior investigation in this area. On the basis of present knowledge, this arterial network is comprised of the ascending branches of the occipital, superficial temporal, supraorbital, and frontal arteries.

Large numbers of branches pass out from all of these vessels and flow together with branches from neighboring arteries to form a rich network. Particularly noticeable in this arrangement is the finding that the longer diameter of this irregular quadrilateral space is usually oriented parallel to the median line. In the region of the forehead the arterial direction is thus oriented diagonally in a superoposterior course, in the region of the vertex it is largely horizontal, and it is anterosuperiorly oriented in the occipital region. If one imagines an elastic network corresponding to a plane in its resting position that is then stretched over the skull by a two-dimensional pull in front and back, one can thus visualize the arrangement for this network. In the corresponding diagrams of anatomic atlases, this relationship is not clearly expressed. Therefore the accompanying drawing [Figure 8, Plate V] was prepared in which the arrangement of the network is clearly visible. It gives the exact picture of a preparation in the anatomic museum at Strassburg, which was prepared by Professor Doctor Pessel. At a sharp angle against the main diameter of the network, the main trunks, namely the rami of the superficial temporal and occipital arteries, penetrate this network, exhibiting a highly tortuous course. Ordinarily, lateral branches originate from the convex tips of the convolutions, those that travel anteriorly coming from the tip of the forward-oriented convolutions.

107. See Figure 9 (Plate V).

Chapter IV

Cutaneous Arteries
of the Upper Extremity
and Nuchal Region

(See Figure 1, Plate I; Figures 10–13, Plate VI)

We begin the description of the cutaneous arteries of the upper extremity with the nuchal region, which was deferred from the previous discussions. This final portion of the trunk corresponds to the upper portion of the trapezius muscle, lying above a line extending from the seventh cervical vertebra to the triangular origin of the scapular spine. Within this area, three territories can be distinguished: a superior one, a lateral one, and an inferior one; however, they cannot be sharply delimited. Their boundaries vary according to the distribution of the participating arteries and are subject to significant variations from case to case.

In general it may be said that the superior territory, associated with the occipital artery, extends as far as the edge of the mandible. The lateral territory, associated with the superficial cervical artery, merges inseparably into the inferior territory supplied by the transverse cervical artery and the transverse scapular artery with numerous anastomoses.

Also included in the superior territory is a portion of the superolateral neck region that was not described previously. This is the skin surface overlying the upper third of the sternocleidomastoid muscle and the gap between this muscle and the anterior border of the trapezius muscle. Along with the superior neck region it is supplied by the superficial descending cervical arteries that originate from the horizontal portion of the occipital artery.[108]

108. Henle, p. 98.

Arnold[109] describes a single superficial descending cervical artery. As a rule, however, there are more, usually three to five (author). They usually penetrate the musculature close to the origin and descend in a vertical direction into the subcutaneous tissue. In some cases they reach the skin surface only at some distance from their origin.[110]

One or two cutaneous branches are regularly found in the upper third of the sternocleidomastoid muscle, emerging from this muscle 3–4 cm inferior to the occipital artery. The most important of these is usually a cutaneous artery arising from the occipital artery in the superior portion of skin overlying the trapezius muscle. This cutaneous artery descends in a course running parallel to the posterior median line. Its medial branches anastomose with symmetric branches from the opposite side in the occipital area and upper nuchal region. Its lateral branches anastomose with branches of the superficial cervical artery.

The superficial cervical artery, supplying the lateral nuchal territory, sends branches to the skin in a twofold manner. One or two cutaneous branches arise directly from the ascending branches that supply the superior portion of inferolateral neck region overlying the lateral edge of the trapezius muscle. These branches arborize at the lateral border areas of this region. The main trunk of the superficial cervical artery[111] courses deep to the middle portion of the trapezius muscle and divides there into several ascending and descending branches that supply primarily the muscle itself. The second group of cutaneous arteries of the middle nuchal territory originates from these branches.

According to the author's observation, these branches penetrate the muscle in varying numbers (three to five) at exit points that are inconstant and course generally parallel to the fibers of the trapezius muscle running toward the scapular spine and thus traveling in a direction opposite to that of their trunk. They usually advance further in the lateral portion of the region. Occasionally one can trace an especially strongly developed branch up to the superficial acromial rete (author). Stahel,[112] who has recently investigated the cutaneous arteries of the shoulder from another point of view, even presents this as a rule. The cutaneous branches of the superficial cervical artery anastomose superi-

109. Arnold, p. 454.
110. In one case investigated by the author, a very unusual arrangement of the descending cutaneous branches from the occipital artery was found. They could be divided into two groups, a superior and an inferior one. The superior group was formed by three small cutaneous branches, which descended to a lesser extent over the sternocleidomastoid, splenius, and trapezius muscles. The inferior group, also comprised of three branches, was noteworthy for the fact that its branches originated not directly from the occipital artery but rather from a common lateral branch. This lateral branch arose from the proximal portion of the occipital artery while still deep to the parotid gland, and then coursed deep to the sternocleidomastoid muscle toward the posterior median line in a slightly descending direction. It terminated in several cutaneous branches at the lateral edge of the trapezius muscle and anastomosed with cutaneous branches of the superficial cervical artery. One cutaneous branch penetrated the sternocleidomastoid muscle, and a second one emerged to the surface over the splenius.
111. Henle, p. 133; Theile, p. 121.
112. Stahel, p. 214.

orly with those of the occipital artery and inferiorly with those of the inferior territory.

This inferior territory, whose entire superior boundary region exists in a reciprocal relationship with that of the middle territory, regularly includes the region of skin overlying the supraspinous fossa inferiorly as well. The cutaneous arteries of the inferior territory originate in part from the transverse scapular artery and in part from the transverse cervical artery. It is associated with the former laterally and with the latter medially. The participation of these arteries cannot be delineated further because of the existence of a widely varying reciprocity between the relative expansion of these two vessels and their numerous anastomoses.

After the transverse cervical artery has reached the superomedial edge of the scapula, it trifurcates into an ascending branch, a descending branch, and a supraspinous branch.

The ascending branch, varying greatly in size, supplies the region of skin overlying the supraspinous fossa.

The cutaneous branches in the region of the previously described dorsal rami of the intercostal arteries arise from the descending branch of the transverse cervical artery.[113]

The supraspinous branch of the transverse cervical artery courses into the supraspinatus muscle to anastomose with the transverse scapular.

The transverse scapular artery courses to the superolateral edge of the scapula. It also gives rise to musculocutaneous branches to the inferior portion of the trapezius. The relative size of these musculocutaneous branches depends on the degree of development of the superficial cervical artery. After giving rise to an acromial branch, the transverse scapular artery courses into the supraspinatus muscle. Its terminal arborization disappears in the deep layers of the infraspinous fossa.

There is little to be said concerning the cutaneous branches of the inferior nuchal territory. Their number varies, and their exit points are not governed by any particular organization. It seems to be a general rule, however, that they course toward the scapular spine as a principal orientation. The cutaneous arteries of the medial region anastomose over the scapular spine with the cutaneous arteries of the infraspinous fossa, whereas the lateral ramifications form an arterial network over the acromion and the insertions of the trapezius and deltoid muscles. In addition, the thoracoacromial and posterior humeral circumflex arteries participate in the formation of this network (*see below*).

Precise information concerning the arterial supply to the region of skin overlying the infraspinous fossa cannot be found in the anatomic literature. Only in the work of Arnold and Theile[114] is a general note given that cutaneous branches arise from the circumflex scapular artery.

---

113. Theile, p. 129.
114. Arnold, p. 490; Theile, p. 134.

According to the author's investigations, a relatively large skin branch, alone providing the arterial supply of this region, is usually given off in the vicinity of the bifurcation of the subscapular artery into the thoracodorsal and circumflex scapular arteries (at the lateral edge of the scapula).

With regard to the position of the origin, some inconsequential variations may be observed.

In the majority of cases this cutaneous artery arises from the circumflex scapular artery shortly beyond the bifurcation point. Sometimes its origin has moved so close to the bifurcation point that it appears to be a third weaker branch of the subscapular artery itself. In a single case, the author noted it to originate from the thoracodorsal artery, 1 cm distal to the origin of the circumflex scapular artery. (In this clearly rare case a smaller cutaneous branch arose simultaneously from the bifurcation of the subscapular artery and supplied a limited cutaneous territory overlying the teres major muscle as far as the boundary of the deltoid.)

In the gap between the teres major and teres minor muscles, this cutaneous artery ascends toward the surface, penetrating the common fascia that covers the muscles of the subscapular space. It wraps around the lateral edge of the teres minor muscle and courses toward the median line in a horizontal direction within the subcutaneous tissue. After crossing the junction between the teres minor and infraspinatus muscle,[115] it divides into three principal branches that diverge toward the three corners of the infraspinous fossa and that give rise to numerous richly arborizing side branches along their courses.[116] The first branch ascends vertically in the general direction of the acromion. Its terminal ramifications course into the portion of the deltoid muscle arising from the scapular spine. These ramifications form part of the superficial acromial rete that was previously mentioned. The middle branch spreads out richly into the triangle defined by a line connecting the center of the scapular spine with the inferior angle of the scapula.

It is thus clear that these ramifications must spread toward the median onto a portion of the trapezius muscle. Superiorly, they anastomose with cutaneous branches from the transverse cervical and transverse scapular arteries. The third branch descends to the inferior corner of the scapula. It courses to the skin overlying the superior edge of the latissimus dorsi muscle which in turn covers the inferior scapular corner. Occasionally (Figure 1, Plate I), yet a fourth branch arises from the bifurcation point. It is located laterally to the third branch and spreads out into the skin overlying the teres major muscle and the bordering areas of the latissimus dorsi muscle.

With its ramifications, this large cutaneous artery, which has just been described in great detail, supplies the skin overlying the infraspinous fossa all by itself. Considerable similarity of arrangement and organization is unmistakable

---

115. In other rare cases, while still over the region of the teres minor muscle.
116. Figure 7 (Plate I): ssa.

between it and the circumflex scapular artery. Perhaps because of this, the name superficial circumflex scapular artery would not be inappropriate.

The shoulder region itself (see Figures 10 and 11, Plate VI) extends as far as the free extremity. The skin of the shoulder area is to be understood here as that cutaneous territory corresponding to the deltoid muscle. It again is divided into a posterior and an anterior region. The border between the two is approximately defined by a line connecting the tip of the acromion with the deltoid tuberosity of the humerus.

The posterior shoulder region, insofar as it is not associated with the superficial circumflex scapular artery (*see above*), is supplied by the posterior humeral circumflex artery.[117] This artery originates at the same level as the anterior circumflex humeral artery.[118] A short distance beyond its origin it ramifies into several terminal branches that primarily supply the deltoid muscle (author). Its terminal ramifications perforate this muscle and supply the subcutaneous tissue of the shoulder area. Above all, however, a very special cutaneous branch (subcutaneous posterior deltoid artery) is of particular importance to this region (author).

It emerges from the posterior humeral circumflex artery between the teres minor muscle and the triceps muscle. Initially, it runs diagonally in a superolateral direction and then wraps itself around the posterior edge of the deltoid muscle. Sometimes the artery shortcuts by perforating the edge of the muscle.

On reaching the subcutaneous tissue overlying the deltoid muscle, it courses toward the acromion (Figure 10, Plate VI). Along this course it gives rise to many posterior and anterior branches. The posterior branches anastomose with branches from the superficial circumflex scapular artery, and the anterior ones arborize over the entire remaining posterior region. One cutaneous branch, which arises shortly beyond the point where the posterior subcutaneous deltoid artery wraps around the posterior edge of the deltoid muscle, is usually larger than the others. It descends parallel to the posterior edge of this muscle and at a short distance from it as far as the insertion on the deltoid tuberosity. In this manner, the entire posterior territory of the shoulder region is supplied by this cutaneous artery. In contrast, the remaining small penetrating branches (two to three) from the posterior humeral circumflex artery play only a minor role. Thus this artery can justifiably be named the posterior subcutaneous deltoid artery.

Superiorly, the ramifications of the posterior subcutaneous deltoid artery anastomose with the superficial acromial rete (*see above*). Along the entire anterior border area the ramifications anastomose with those of the anterior subcutaneous deltoid artery of the anterior region.

As a rule the anterior shoulder region is also principally supplied by an especially strongly developed cutaneous artery (Figure 11, Plate VI). This artery

---

117. The cutaneous relationships of this artery are mentioned in greater detail only by Stahel.

118. Henle, p. 138.

is associated with the thoracoacromial artery. At the superior edge of the pectoralis minor muscle this artery divides into its diverging terminal branches. The relationships of the pectoralis branches to the anterior skin of the breast have already been discussed. The deltoid branch and the acromial branch are important to the anterior shoulder area.

The deltoid branch[119] descends in the deltopectoral groove. To a minor extent, its branches pass to the lateral border of the pectoralis major muscle, but their principal distribution is to the clavicular portion of the deltoid muscle. Approximately in the middle of the deltopectoral groove or slightly superior to it, a cutaneous artery (anterior subcutaneous deltoid artery) emerges from the deltoid branch, penetrates the superficial fascia, and curves horizontally toward the anterior surface of the deltoid muscle.[120] It soon ramifies into diverging, abundantly branched terminal ramifications. Some of these ascend to the clavicle, others continue in the direction of the trunk, and descending branches spread out in the direction of the deltoid tuberosity (anastomose with the posterior subcutaneous deltoid artery and with the superficial acromial rete; *see above*).

In the anterior region as well, individual (two to three) small, perforating cutaneous arteries from the deltoid branch play a subordinate role.

The anterior subcutaneous deltoid artery is usually present but not with the same consistency as the posterior subcutaneous deltoid artery.

Sometimes the anterior subcutaneous deltoid artery is replaced by two or three cutaneous arteries that emerge at a considerable distance from one another from the deltopectoral groove. The most superior one arises from the origin of the deltoid branch. The middle one emerges at the lower boundary of the upper third, whereas the inferior one arises in the middle of the deltopectoral groove. In this case, numerous important cutaneous branches also pass into the skin overlying the lateral border of the pectoralis major muscle.

In other cases, the normally small perforating cutaneous branches are strongly developed and supply the usual territory of this artery in a pattern corresponding to one of the branches of the anterior subcutaneous deltoid artery.

The acromial branch of the thoracoacromial artery also participates in supplying the skin of the shoulder region. This branch courses laterally parallel to the clavicle. Inferior to the clavicular insertion of the deltoid muscle, it passes to the acromion between the clavicle and the coracoid process. "It terminates in branches which, having penetrated the appendage of the deltoid muscle, form a network around the acromioclavicular joint together with the acromial branch of the transverse scapular artery." (Henle[121]).

The relationship between the acromial branch of the thoracoacromial artery and the acromial branch of the transverse scapular artery has recently been described in greater detail by Stahel.[122] It consists of a ring-shaped arterial arcade that surrounds the acromioclavicular part of the deltoid and the acromion

119. Henle, p. 137.
120. The author.
121. Henle, p. 138.
122. Stahel, p. 213.

itself just inferior to the origin of the deltoid muscle. Numerous branches course to the acromial rete from this arcade.

The acromial rete encloses the surface of the acromioclavicular joint.

It consists of a deep and a superficial layer. The deeper layer is associated with the periosteum of the acromioclavicular joint, while the superficial layer (superficial acromial rete) is contained within the subcutaneous tissue and extends onto the neighboring insertion edges of the trapezius and deltoid muscles.

The acromial rete is distinctive in the diversity of sources from which it is formed, as a summary of them will demonstrate. The following could be traced to the acromial rete:

1. Branches of the transverse scapular artery
   a. From the subcutaneous supraclavicular artery (see Chapter II)
   b. From the acromial branch, which originates from the main artery shortly before it enters the supraspinous fossa
   c. Penetrating skin muscle branches of the lateral portion of the supraspinous fossa of the scapula
2. A cutaneous branch (a constant one, according to Stahel) from the superficial cervical artery
3. Cutaneous branches of the medial portion of the supraspinous fossa from the transverse cervical artery
4. Terminal branches of one partial branch of the superficial circumflex scapular artery which ascends to the superolateral corner of the infraspinous fossa from the subscapular artery
5. The ascending branches of the posterior subcutaneous deltoid artery from the posterior humeral circumflex artery
6. Cutaneous branches from the thoracoacromial artery
   a. From the deltoid branch, anterior subcutaneous deltoid artery
   b. From the acromial branch

The upper arm region is joined to the shoulder region (see Figures 10 and 11, Plate VI). Until now, this area has been described only in very vague terms in the anatomic literature. According to the literature, it is supplied by branches of the brachial artery and its collateral arteries. These vessels send out numerous unnamed secondary branches that arborize within the region.[123] They are so variable that they do not even exhibit similarity within the two extremities of the same individual.[124] This rather inconclusive summary of that information found in existing handbooks suggests an admonition against too specialized a description, the likes of which follows. In fact, as is the case in other regions, if one would like to describe individual cutaneous arteries, one could describe just as many variations as there are cases investigated. On the other hand, many principles emphasized during the course of this presentation

123. Sappey, p. 632.
124. Henle, p. 141.

are further confirmed in the upper arm. As variable as these cutaneous arteries may be in their origins and in the number of small branches arising from the deeper vessels, there is yet a great regularity to be found in their region of arborization, in their arrangement, and in the main direction of their courses in all cases. Considerable emphasis must be placed on this.

Gegenbaur, at any rate, makes a general attempt at taking these relationships into consideration. He divides the branches of the brachial artery in the upper arm into those that fall on the flexor side and those that fall on the extensor side. "The former often arise from the trunk which is already turned toward the flexor side. They supply the muscles and skin of the flexor side. Those that go to the extensor side are usually branches of a single stronger branch, the deep brachial artery. Individual branches from both groups are distributed to the extensor side of the elbow joint and enter the articular rete which is to be found there as well as at the epicondyles (collateral arteries)."[125]

This arrangement shows that Gegenbaur had the muscular branches in mind exclusively when he formulated it. The separation of the brachial branches into direct branches of the brachial artery and those branches of the deep brachial artery, both of which supposedly supply separate territories, has validity only for the muscular branches but not for the cutaneous branches.

In the author's investigation the following distributional characteristics became apparent:

The lateral half of the upper arm territory is supplied for the most part by the inferior radial collateral artery.[126] This artery is the superficial continuation of the deep brachial artery, which descends between the long and medial triceps bellies, and which divides into its terminal branches at approximately the level of the deltoid tuberosity of the humerus.[127] Deep to the lower edge of the origin of the lateral triceps belly (or passing through it; author), the radial collateral artery emerges to the surface of the arm in the fissure between this muscle and the medial triceps belly, about the midpoint between the acromion and the lateral epicondyle.[128] While within this fissure, it passes directly deep to the superficial fascia of the arm. Slightly above the radial epicondyle it penetrates the superficial fascia, approximately at the same position as the posterior inferior cutaneous branch of the radial nerve, and then passes with its terminal branches into the more superficial position of the cubital arterial rete which will be discussed later. While passing toward the surface, it gives rise to anterior and posterior cutaneous branches that penetrate the superficial fascia of the arm overlying the muscle fissure. Sometimes an anterior and a posterior branch emerge from a short common stem. The number of these branches varies, as has already been pointed out above.

125. Gegenbaur, p. 691.

126. Inferior radial collateral artery, Henle. External collateral artery. Superficial branch of the profunda brachii artery.

127. Henle, p. 142.

128. See Figure 10 (Plate VI): 2.

The posterior branches, usually four to six in number, supply the entire posterior portion of the lateral upper arm territory with the single exception of a superior triangle defined by drawing a horizontal line through the insertion of the deltoid muscle on the deltoid tuberosity up to the posterior border of the region (*see below*).

All posterior branches pass in a horizontal direction toward the posterior boundary of the region, where they anastomose with similar branches from the medial upper arm surface. With their abundant ramifications, they supply the skin overlying the lateral surface of the lateral triceps belly. It is important to emphasize that other cutaneous branches, for example, those from the muscular branches of the lateral triceps belly, were not observed in this region. Through its descending branches, the inferior posterior branch anastomoses with the network arising on the surface of the elbow.

The superiormost posterior branch is worthy of note. Since it supplies the superior portion of the territory, in which the radial collateral artery is not yet superficial, it usually exhibits a characteristic course. It turns superiorward from its point of departure in a sharp curve and ascends parallel to the radial collateral artery in an opposite direction as far as the deltoid tuberosity. The horizontal branches then pass out from this stem.

The superior triangle of the posterior territory of the lateral surface, already noted above, is associated with the posterior humeral circumflex artery.[129] One to three of its branches emerge inferior to the posterior edge of the deltoid muscle and travel horizontally in a posterior direction, sending out numerous secondary branches along the way. The most inferior of these branches anastomoses with branches from the uppermost posterior branch of the radial collateral artery. According to Theile, perforating branches from the triceps muscular branches, those originating from the deep brachial artery, also course to the surface here. They anastomose with the cutaneous branches from the posterior humeral circumflex artery.

The anterior cutaneous branches of the radial collateral artery are less numerous (three to four as a rule). They are associated with that cutaneous territory of the lateral upper arm region that corresponds to the brachialis muscle, the origin of the brachioradialis muscle, and the central portion of the biceps.[130] The superior and inferior portions of the skin surface overlying the lateral biceps receive their cutaneous arteries from the brachial artery.[131]

The superior of these anterior branches from the radial collateral artery exhibit a horizontal course. With abundant branchings, they supply the upper medial brachialis and middle biceps regions (and anastomose with branches of the medial region). In contrast, the inferior branches pass inferiorward to the upper forearm, descending diagonally in an anterior direction, and come into

129. Figure 10 (Plate VI): 1.
130. For this territory, associated with Gegenbaur's flexor region, their distribution is thus not correct.
131. See Figure 10 (Plate VI): 3.

anastomotic communication there with cutaneous branches of the forearm arteries.

The inferolateral portion of the biceps region is associated with a direct cutaneous artery from the brachial artery.[132] It emerges from the brachial artery approximately in the center of the distance between the deltoid tuberosity and the lateral epicondyle. In the groove between the biceps and brachialis muscles, it ascends to the surface, perforates through the superficial fascia about the level of the lateral epicondyle, and then makes anastomotic connection with the recurrent radial artery. A few cutaneous branches emerge from it and pass in a weakly ascending or horizontal anteriorward course to the skin overlying the lower lateral biceps surface in anastomotic connection with branches of the medial region. The superior third of the cutaneous surface overlying the lateral biceps is associated with cutaneous branches of the medial upper arm and thus leads us to this region.

The medial upper arm region can also be divided into an anterior and posterior territory. The boundary between the two is formed by the brachial artery itself, descending in the groove between the flexors and extensors of the upper arm in a rather superficial position.

Gegenbaur's division holds true for the anterior territory. It is supplied without exception by direct cutaneous branches of the brachial artery. It corresponds to the medial surface of the coracobrachialis and biceps muscles. Six to eight or more branches of the artery penetrate the superficial fascia directly over the brachial artery and travel in a horizontal or weakly descending direction toward the anterior boundary of the region, giving rise to numerous ramifications along the way. The superiormost of these arteries encroaches over the boundary line, as already noted, and also supplies the superolateral territory of the biceps muscle in front of the anterior edge of the deltoid muscle.[133] The middle branches are in anastomotic contact with anterior branches of the radial collateral artery, whereas the inferior ones anastomose with branches arising from the cutaneous artery which emerges between the brachialis and biceps muscles (*see above*).

In contrast, this arrangement does not hold true in any way for the posterior territory, since it is supplied in part by direct cutaneous branches from the brachial artery and in part by cutaneous branches arising from the superior ulnar collateral artery, which is only rarely a branch of the deep brachial artery.[134] At the same level as the superiormost branch of the anterior territory, a number of smaller arteries arise from the brachial artery and pass to the lymphatic glands and that portion of the axilla belonging to the upper arm.[135]

The territory of the superior ulnar collateral artery anastomoses with these branches. The superior ulnar collateral artery arises about the same level

132. See Figure 10 (Plate VI): 4.
133. See Figure 11 (Plate VI): 4.
134. According to Quain, 35 times in 500 cases (compare Henle, p. 280).
135. Several of the same are often united to form a common trunk, or they may emerge from the uppermost branch of the anterior territory. See Figure 11 (Plate VI): 4.

or slightly inferior to the origin of the deep brachial artery from the brachial artery.[136] After a short course it divides into a number of descending branches. The majority of these are muscular branches and terminate in the medial triceps belly.

On the other hand, one branch, the superficial ramus of the superior ulnar collateral artery, is a prominent cutaneous arterial trunk. It is not unusual for it to arise independently[137] from the brachial artery and is often of significant diameter (the author).

It descends in the fissure of the brachialis muscle, and in the case of a higher origin additionally in the groove between the long and medial triceps bellies directly deep to the superficial fascia, and passes into the superficial layer of the cubital rete with its terminal branches (anastomoses with the recurrent ulnar artery). Arising from it are anterior and posterior cutaneous branches which perforate the superficial fascia. All of them exhibit a weakly descending or nearly horizontal course.

The anterior branches supply the skin as far as the brachial artery and anastomose with the small, direct cutaneous branches coming from it that course posteriorly.

The stronger posterior branches arborize as far as the posterior cutaneous boundary overlying the medial surface of the long triceps belly. They make anastomotic contact with the posterior branches of the radial collateral artery on the posterior surface of this muscle.

In the superior portion of the region, a few perforating musculocutaneous branches from the superior ulnar collateral artery and the deep brachial artery are also found.[138] Their expansion is reciprocal with that of the superior branches of the superficial ramus.

The cutaneous surface overlying the posterior elbow joint is joined to the upper arm region. It is in a way the connecting link between the upper arm and the forearm and belongs in some respects to both. With it, we leave a region that essentially has not been described until now and into a well-known and fully investigated territory for which a short review of the established facts is sufficient. In the posterior elbow region, an extensive vascular network is formed by the arteries of the upper arm and the forearm. Within it a superficial and a deep layer, which have significant interconnection, can be differentiated. The superficial layer is associated with the skin and subcutaneous tissue, whereas the deep layer is associated with the muscles and joint capsule.

The principal arrangement of this arterial network has been well described by Henle[139] : "Three main branches are distinguished in the cubital network, two vertical ones to the sides of the olecranon and one transverse one above it. The radial vertical branch, lying in the hollow between the radial head and

136. Henle, p. 142.
137. Henle, p. 143. See Figure 11 (Plate VI): 3.
138. Figure 11 (Plate VI): 16.
139. Henle, p. 152.

the olecranon, is created by the confluence of the inferior radial collateral artery and the recurrent interosseous artery. The anastomosis of the superior ulnar collateral artery with the posterior branch of the recurrent ulnar artery forms the ulnar vertical branch. The lateral arcade consists of the interconnection of branches from the inferior radial collateral artery with the superior ulnar collateral artery overlying the olecranon." Arising from these arteries are numerous cutaneous vessels which comprise the superficial layer of the cubital rete in the subcutaneous tissue. Within this layer, no particular arterial orientation predominates.

The remaining region of forearm skin[140] is divided into two territories, that of the radial artery and that of the ulnar artery.

The cutaneous territory of the radial artery corresponds to the radial side and a smaller portion of the flexor side of the forearm. The remainder of the forearm skin receives its arterial supply from the ulnar artery. The boundary between the ulnar and radial territories is denoted on the extensor side by the groove between the extensor carpi radialis brevis and extensor digitorum communis muscles, that is to say the abductor pollicis longus muscle. On the flexor side the border line extends distally from the biceps insertion approximately overlying the fissure between the brachioradialis and pronator teres muscle. From approximately the boundary of the upper and middle thirds of the forearm, it curves ulnarward and defines the territory of skin overlying the smaller distal portion of the pronator teres, flexor carpi radialis, and palmaris longus muscles and over the aponeurotic surface on the tendons of these muscles to the radial artery (see Figure A, Plate IX). Following the superficial edge of the flexor carpi ulnaris muscle, it then descends to the wrist.

The radial territory is traversed lengthwise by the radial artery. Along its course the artery gives rise to two groups of branches that connect with the skin, passing both radialward and ulnarward. The superiormost of the radially oriented branches is distinguished by exceptional size and constancy. It has been named the recurrent radial artery.[141] This artery emerges from the radial artery in a horizontal or weakly descending direction and then curves superiorly in a hook-like fashion. From the convexity of the arch thus formed, branches originate that pass to the supinators and the muscles of the radial edge (extensor carpi radialis longus and brevis). In the vicinity of the lateral epicondyles, the artery splits into several terminal branches that pass in part to the supinator muscle, in part to the brachialis, and in part to the skin of this region. These cutaneous branches anastomose with terminal ramifications of the lowest anterior branch of the radial collateral artery and with branches of that cutaneous artery that emerges from the groove between the biceps and brachialis muscles and nourishes the inferior portion of the lateral biceps cutaneous surface. The branches that arise from the convexity of the arch also pass to the skin through varied connections. A larger cutaneous branch usually emerges from the supinator

140. For this region, reliable information from the anatomic literature is incomplete. We therefore present our own observations. Compare Figures 13 and 14 (Plates VI and VII).

141. Henle, p. 145; Sappey, p. 636.

muscle and spreads out with numerous ascending and descending branches in the subcutaneous connective tissue overlying the fascia of this muscle. A variable number of cutaneous branches also emerge from the two fissures formed by the brachioradialis, extensor carpi radialis longus, and extensor carpi radialis brevis muscles. They perforate the forearm fascia and supply the corresponding cutaneous territory. These arteries also anastomose with branches of the inferiormost anterior branch of the inferior radial collateral artery. In addition, they anastomose with the superficial cubital rete and the neighboring boundary territory of the ulnar region.

The remaining radially oriented branches of the radial artery are in part also musculocutaneous branches (brachioradialis muscle) and in part (in the inferior half of the territory) true direct cutaneous branches. Immediately after their departure from the radial artery, they penetrate the forearm fascia and arborize in a diagonally descending direction to the boundary of the region in the subcutaneous connective tissue. The cutaneous branches of the radial artery that are directed toward the ulnar are numerous only in the inferior portion of the territory. They move toward the ulnar boundary of the region following a weakly descending course.

The ulnar territory is traversed lengthwise by the ulnar artery.

Along its entire course this artery gives rise to numerous branches that supply the muscles and skin of the ulnar region. One portion of the ulnar territory lies on the flexor side of the forearm. Its radial boundary is defined by the border that was extensively described above. On its ulnar side, it reaches the anterior edge of the flexor carpi ulnaris muscle.

Of the branches of the ulnar artery, the anterior recurrent ulnar artery and the anterior interosseous artery are consistent in their participation in the cutaneous supply of this region.

The anterior recurrent ulnar artery[142] originates from the ulnar artery at the same level as the posterior recurrent ulnar artery. In its initial portion, it is often joined with the latter in a short common stem. According to Henle, this situation is even the norm.[143] In the groove between the brachialis and pronator teres muscles, it ascends to the medial epicondyle and anastomoses there with terminal branches of the superior ulnar collateral artery. A number of cutaneous branches arise from it, some penetrating the pronator teres muscle and some passing through the groove formed by this muscle and the flexor carpi radialis to the skin of the flexor side of the elbow joint. Their arborization there exists in reciprocal relationship with those cutaneous branches coming out of the terminal portion of the brachial artery and spreading into the ulnar region. One of these occurs especially frequently. It has been described by Gruber[144] under the name superficial cubital fold artery. It originates from the

142. Sappey, p. 640.

143. Henle, p. 149.

144. Zeitshrift der Gesellschaft der Aerzte zu Wien. Jahrg. VIII, Bd. II, p. 481. Compare Henle, p. 143.

terminal portion of the brachial artery, more rarely from the inferior ulnar collateral artery or from the radial artery superior or posterior to the superficial tendon of the biceps muscle. Beneath the fascia of the forearm it courses diagonally toward the median and then terminates in musculocutaneous branches in the groove between the flexor carpi ulnaris and palmaris longus muscles. The cutaneous branches supply the ulnar edge of the territory. When this artery originated at a higher location, the author often observed two or three additional small direct cutaneous branches from the terminal section of the brachial artery that arborized in the skin overlying the superior portion of the pronator teres and flexor carpi radialis muscles.[145] All of these cutaneous branches are in contact with one another.

The remaining portion of the ulnar arterial territory on the flexor surface of the forearm belongs to constant cutaneous branches from the median artery.[146] This vessel arises from the superior portion of the anterior interosseous artery or even from the common stem of the two interosseous arteries. Situated between the flexor digitorum profundus and sublimis muscles, it passes distally anterior to the median nerve that accompanies it. Its size is exceptionally variable. Usually it is weak and terminates in a few musculocutaneous branches of this region. On the other hand, it is sometimes imposing in size and can be traced into the palm of the hand. Its cutaneous branches sometimes appear in the groove between the flexor carpi radialis and palmaris longus muscles and sometimes between the palmaris longus and the flexor carpi ulnaris muscles. They pass radially in a weakly descending direction.

The extensor portion of the ulnar arterial territory is principally associated with the two interosseous arteries. The anterior interosseous artery supplies only the small triangular cutaneous territory corresponding to the superficial portion of the abductor pollicis longus, extensor pollicis brevis, and extensor pollicis longus muscles. Everything else belongs to the posterior interosseous artery. After this artery has given rise to the recurrent interosseous artery to the superficial cubital rete, it descends to the wrist in the groove between the ulnar muscles. Along its course it gives rise to a large number of muscular branches to neighboring muscles. The cutaneous branches originate from these muscular branches and in the inferior portion of the region probably also directly from the main arterial trunk. One can differentiate two groups of cutaneous branches. The cutaneous arteries of one group appear between the flexor carpi ulnaris and extensor carpi ulnaris and those of the other group between the extensor carpi ulnaris and the extensor digitorum communis. In both groups there are branches passing ulnarward and radialward. All demonstrate a more or less diagonally descending course. With their ramifications, they supply the entire region. Only in the region of the upper lateral portion of the flexor carpi ulnaris do a few vertically

---

145. Figure 12 (Plate VI): a–d.

146. According to Sappey, the median artery is constant in its occurrence. According to Gruber, it was absent twice in 100 arms. Zeitschrift der Gesellschaft der Aerzte zu Wien. 1852 II 492. Compare Henle, p. 151; Sappey, p. 641.

descending cutaneous branches, originating from musculocutaneous branches of the ulnar artery, perforate the forearm fascia.

It has already been noted that the portion of the ulnar region corresponding to the superficial portions of the abductor pollicis longus, extensor pollicis brevis, and extensor pollicis longus muscles is associated with the anterior interosseous artery. Posteriorly, this artery gives rise to three or four perforating interosseous branches that pass through the interosseous ligament to the extensor side of the forearm, distributing themselves among these three muscles and the extensor indicis proprius.[147] As a rule three cutaneous arteries emerge from them. The most proximal of these passes inferiorly at the upper edge of the abductor pollicis longus muscle, the middle one descends in the anatomic snuffbox, and the inferior one passes along the edge of the extensor pollicis brevis. All three arborize into numerous cutaneous branches that penetrate the superficial fascia. They form an arterial network in the subcutaneous connective tissue of the territory which is in anastomotic connection with the superficial vascular network of the wrist.

In this manner, we pass to the skin of the hand and thereby are again in a region that is so well known and thoroughly described that reference to the anatomic literature will suffice.

147. Sappey, p. 641. According to Henle (p. 151), five to six perforating interosseous branches.

Chapter V
Cutaneous Arteries of the Lower Extremity

(See Figure 5, Plate III; Figures 14 and 15,
Plate VII; Figures 16, 17, and 18, Plate VIII)

To conclude the description, only the cutaneous arteries of the lower extremities remain. We begin with the skin of the gluteal region. This is that region of skin corresponding to the gluteus maximus muscle and the anterior superficial portion of the gluteus medius muscle. The branches of the hypogastric artery, which supply the muscles of the buttocks, also give rise to the cutaneous arteries of this area. The region is divided into several territories, each belonging to one or another of the branches of this main artery. The principal branches are the superior and inferior gluteal arteries and secondarily the lateral sacral, the internal pudendal, the iliolumbar, and the superficial circumflex iliac arteries. Those arteries mentioned secondarily are, along with their cutaneous branches, associated with the trunk, but their terminal ramifications spread to the entire superomedial edge of the region; they will be discussed later. The boundary between the territories of the superior and inferior gluteal arteries is approximated by a line that connects the inferior tip of the sacrum with the greater trochanter. As there is some variability in this boundary, it will serve only for general orientation. In individual cases it is often encroached from one side or the other.

Following its exit from the pelvic cavity, the superior gluteal artery divides into a deep and a superficial branch. The superficial branch is situated between the gluteus maximus and medius muscles. It courses anteriorly between these two and branches out in the superior portions of these muscles and in the corresponding cutaneous territory.

The deeper branch courses anteriorly between the gluteus medius and minimus muscles. It soon divides into a superior and inferior branch. The superior branch travels along the insertion edge of the gluteus minimus muscle. It anastomoses with the iliolumbar artery, the superficial circumflex iliac artery, and the lateral circumflex femoral artery. From both branches, a few cutaneous arteries arise that penetrate the fascia in that part of the region bordering on the anterior edge of the gluteus maximus muscle.

The inferior gluteal artery emerges from the pelvic cavity inferior to the piriformis muscle. It divides into an ascending ramus and a descending ramus. The former usually terminates in musculocutaneous branches of the inferior portion of the gluteus maximus muscle, lying below the aforementioned boundary line. Of the branches of the descending ramus, only those that supply the inferomedial edge of the gluteus maximus muscle and its corresponding skin territory are of importance here.

The cutaneous arteries of both territories exhibit few characteristic qualities. They penetrate the thin superficial fascia in varying numbers. Their exit points also do not seem to be governed by any special organizational rules. It can only be said that they are more numerous in the medial portion of the territory than in the lateral portion. However, all cutaneous arteries of the region demonstrate the characteristic of convergence toward the greater trochanter in their principal orientation. The arteries of the territory of the superior gluteal artery therefore exhibit a diagonal inferolateral direction, which becomes steeper in the more lateral portions of the territory. The arteries of the inferior territory (of the inferior gluteal artery) course lateralward in a horizontal direction, and those of the lower medial border region even course in a weakly ascending direction. Above the greater trochanter, a few cutaneous arteries from the inferior territory anastomose with a few from the superior territory. However, this does not result in the formation of a true arterial network, as is the case in the shoulder region, for example.

The superomedial border region up to the anterior superior iliac spine is supplied by arteries of the trunk, as was already indicated.

The inferior portion of the medial edge, extending from approximately the inferior end of the sacrum to the point where the inferomedial edge of the gluteus maximus muscle leaves the common muscle belly of the semitendinosus and biceps femoris muscles is supplied by branches of the internal pudendal artery.[148]

These branches emerge in part from muscular branches that the internal pudendal artery gives off to the gluteus maximus muscle in its course outside of the pelvis. They appear on the superficial fascia 1–2 cm from the medial edge of this muscle. In part, however, the internal pudendal artery also gives rise to direct cutaneous branches in the ischiorectal fossa (the author). Sometimes

---

148. Observation of the author. Sappey's statement that this cutaneous territory and the additional cutaneous surface overlying the coccyx and sacrum is supplied by branches from the ascending branch of the inferior gluteal artery (p. 658) was not confirmed by the author in his dissections.

one of these cutaneous arteries also arises from one of the external hemorrhoidal arteries.

These arteries, two to three in number, course through the adipose tissue of the ischiorectal fossa to the inferomedial edge of the gluteus maximus muscle. They wrap themselves around it and spread out over the superficial fascia in the subcutaneous connective tissue. All cutaneous arteries arising from the internal pudendal artery have the same principal direction as that of the inferior gluteal artery, with which they anastomose in part. Their lateralward extension exists in reciprocal relationship with the inferior gluteal artery. The most superior branches anastomose to one side with cutaneous branches from the lateral sacral artery and to the other side, above the coccyx, with the terminal branches of the external hemorrhoidal arteries.

It has already been pointed out that the cutaneous branches from the lateral sacral artery extend onto the corresponding portion of the edge of insertion of the gluteus maximus muscle.

The entire superior edge of the region is supplied by cutaneous branches from the dorsal ramus of the iliolumbar artery, i.e., the last lumbar artery (refer to the cutaneous arteries of the abdomen). Directly over the iliac crest, they penetrate the fascia in varying numbers (three to five) and descend into the subcutaneous connective tissue above the gluteus medius muscle. They anastomose with ascending cutaneous branches from the superficial circumflex iliac artery.

The posterior femoral region joins the skin of the buttocks.[149] This region extends inferiorly to the popliteal fossa. At the medial and lateral surfaces of the thigh, it is in conjunction with the anterior femoral region without sharply defined boundaries.

The posterior femoral region divides into a superior and an inferior cutaneous territory.

The boundary between the two may be approximated by a line that passes around the posterior femoral surface in a horizontal direction at the level of the lesser trochanter.

Participating in the supply of the upper region are:

1. Branches of the obturator artery
2. Branches of the internal pudendal artery
3. Branches of the inferior gluteal artery
4. Branches of the medial circumflex femoral artery, and
5. Branches of the lateral circumflex femoral artery

Those branches that course out of the obturator artery to the superiormost portion of the medial thigh surface were previously described in connection with the cutaneous arteries of the genital and perineal areas.

---

149. Compare Figure 15 (Plate VII).

Associated with the internal pudendal artery is the medial border territory, which is joined laterally to the perineum (author). Its cutaneous branches usually do not arise from the internal pudendal artery itself, but rather from the perineal artery or from the lateralmost of the scrotal arteries.[150] They turn laterally at the edge of the ischium and arborize in the subcutaneous connective tissue overlying the ischial origin of the adductor magnus muscle as far as the semimembranosus and semitendinosus muscles. In the principal direction of their ramifications, they show a horizontal or slightly descending course. Their lateral expansion exists in reciprocal relationship with the cutaneous branches from the inferior gluteal and medial circumflex femoral arteries.

At the inferomedial edge of the gluteus maximus muscle, a few more cutaneous arteries emerge from the inferior gluteal artery and spread to the posterior femoral region to a limited extent.[151]

Of greater importance are the branches of the medial circumflex femoral artery.[152] Its descending terminal branch alone is involved in supplying the skin of the femoral region. It emerges through the groove between the quadratus femoris and adductor magnus muscles.[153] Terminal branches supply the origins of the superficial flexors and in part also the vastus lateralis muscle. At exit points that are not well defined, cutaneous branches from the medial circumflex femoral artery perforate the fascia of the upper thigh. In a horizontal direction, they course both laterally and medially. Toward the median, they are in anastomotic contact with the branches of the internal pudendal and inferior gluteal arteries, and laterally with branches of the lateral circumflex femoral and superficial circumflex iliac arteries. The lateral circumflex femoral artery will be discussed along with the description of the anterior femoral region.

The inferior territory of the posterior femoral region is primarily associated with branches of the perforating rami from the profunda femoris artery. But the popliteal artery is also an important participant in the supply of this area.

The perforating rami,[154] as is known, course through the fibrous arches of the adductor tendon to the posterior side of the superior thigh. There, as a rule, they divide into three diverging terminal branches. One of these courses laterally in a horizontal direction to the vastus lateralis muscle. The other two, descending and ascending rami, distribute themselves on the long flexors. Cutaneous arteries originate from all three. The largest and most prominent (author) of these ascend to the surface in the groove between the long and short heads of the biceps femoris muscle (four to six in number).[155] They penetrate the femoral fascia. Immediately superficial to the fascia, and sometimes deep to it,

150. Figure 6 (Plate IV).
151. Figure 15 (Plate VII).
152. Henle, p. 192.
153. Sappey, p. 672.
154. Sappey, p. 674; Henle, p. 192.
155. Figure 15 (Plate VII).

they divide into a stronger lateral branch and a weaker medial one. Abundantly ramifying, the lateral branches course over the posterior portion of the vastus lateralis muscle and fascia and pass laterally in a weakly descending direction. At the lateral surface of the thigh, they anastomose with cutaneous arteries of the anterior region. The medial branches also have a weakly descending course.

In addition to this group, which is particularly distinguished by its exit points, a few more cutaneous branches in varying numbers appear at locations on the fascia that are not especially well defined. Their number and the size of their territory of expansion are in inverse proportion to the branches of the former group. In the medial portion of the territory, they appear to be stronger and more numerous than in the lateral region (author).

Definite boundaries between the territories of the individual perforating rami cannot be given. They exist in reciprocal relationship. But since the first and the third perforating rami are generally the stronger ones,[156] the territories corresponding to them are usually also larger than that of the middle perforating ramus.

The inferiormost portion of the inferior territory is supplied by cutaneous branches from the popliteal artery. In the superior part of the popliteal fossa, three to four cutaneous arteries emerge, often united in their origin into a common trunk. A few of these small branches course lateralward at a more or less sharp angle and arborize in the skin overlying the terminal portion of the biceps femoris muscle. Similar medial branches disappear into the subcutaneous connective tissue overlying the tendons of the semitendinosus and semimembranosus muscles. Perhaps worthy of special attention is the following cutaneous artery, which was observed repeatedly by the author. It originates either directly from the popliteal artery or from the common trunk located in the popliteal fossa which was mentioned above. In the groove between the semitendinosus and the long head of the biceps femoris muscle, it ascends directly deep to the fascia. It may reach a substantial length. In one case it could be traced as far as the union of the common heads of the semitendinosus and biceps femoris muscles.[157] Branching off from it are medial and lateral cutaneous arteries, which penetrate the fascia immediately overlying the muscular groove. Their strength and extent are in reciprocal relationship with the medial cutaneous branches from the perforating rami of the profunda femoris artery.

One cutaneous arterial territory of the anterior femoral region[158] has already been described in connection with the cutaneous arteries of the abdomen and of the external genitalia. This territory, associated with the superficial circumflex iliac, inguinal, external pudendal, and obturator arteries, is separated from the rest of the femoral region by a line that extends from the anterior superior iliac spine along the medial edge of the sartorius muscle. This line leaves the

156. Henle, p. 193; Sappey, p. 674.
157. Compare Figure 15 (Plate VII).
158. Compare Figure 4 (Plate II) and Figure 5 (Plate III).

sartorius at the point where the femoral artery courses beneath the muscle and extends medially parallel to the inguinal ligament.

That portion of the anterior femoral area lying below this line[159] divides into two territories. One of these is associated with the musculocutaneous branches of the femoral artery. It encloses the entire medial and inferolateral portions of the region. The superolateral portion is usually associated with branches of the profunda femoris artery. Sharp boundaries cannot be drawn between the two territories for "the relationship of the profunda femoris artery to the main trunk of the femoral artery is so variable that the norm can only be determined with difficulty."[160]

Along its course through Hunter's canal between the vastus medialis and the adductor muscles, the femoral artery gives rise to several musculocutaneous branches that supply the first of the two territories. They are subject to unusual variations in number and exit points. Only the inferiormost one, which is also the largest, is fairly constant in its origin and in the arrangement of its branches. It has been described by various authors under the name supreme articular knee artery,[161] superficial articular knee artery, and large anastomotic artery. Its name indicates that it participates substantially in the formation of the superficial arterial network of the knee.

These branches of the femoral artery may be divided into a medial and a lateral group. Included with the medial group are all branches that penetrate the femoral fascia medial to the medial edge of the sartorius muscle; those branches of the lateral group appear on the lateral edge of this muscle.

Actually the external pudendal arteries also belong to the medial group. All of their branches (four to six in number) share the common feature of a medial course in a horizontal (in the superior portion of the region) or descending direction (in the inferior portion of the region). In so doing they exhibit a more or less highly tortuous course. Arising from them are ascending and descending branches that branch abundantly and distribute themselves throughout the skin of the region.

A very similar situation holds true for the lateral group, whose branches travel lateralward. The same highly tortuous course in a descending direction is also evident here.

The superolateral territory of the femoral region is associated with the profunda femoris artery or, to be more exact, the lateral circumflex femoral artery.[162] Before passing deep to the origin of the rectus femoris muscle, this artery sends out musculocutaneous branches to the anterior portion of the gluteus muscle, to the origins of the sartorius and tensor fascia lata muscles and central portion of the rectus femoris muscle. French anatomists[163] describe these as

159. Compare Figure 14 (Plate VII).
160. Henle, p. 189.
161. Henle, p. 193; Arnold; Sappey; Harrison.
162. Theile, p. 227.
163. Sappey, p. 672.

branches of an independent great superficial muscular artery. They do this with the same privilege as those who designate them as branches of the lateral circumflex femoral artery, for the number of variations in origin is unusually large here. These muscular branches very often arise from the profunda femoris artery itself or from the femoral artery.

The cutaneous arteries of this region penetrate the fascia lata at various locations. Some appear at the medial edge of the sartorius muscle, some come out of the groove between the sartorius and rectus femoris muscles, and others appear in the region of the rectus femoris. The more superior cutaneous arteries of this region exhibit an almost vertically descending or somewhat medially oriented course. The more inferior ones branch abundantly and course in the same laterally descending diagonal arrangement that was described for the lateral branches of the femoral region.

Their terminal ramifications are in anastomotic contact with the cutaneous arteries of the posterior femoral region and superiorly with descending cutaneous branches from the superficial circumflex iliac artery over a considerable range. The small superficial muscular artery, which Sappey[164] describes for this region, appears to correspond essentially to those branches. Only the origin is described differently. "The small superficial muscular artery," states Sappey, "arises from the superiormost part of the femoral artery, at the same level or slightly below the superficial inferior epigastric artery, sometimes sharing a common initial stem with the latter. It courses lateralward and gives rise (in addition to muscular branches) to branches in the skin overlying the tensor fascia lata and the initial portion of the sartorius muscle." The terminal expansion of the arteries of this region shows greater uniformity than the arterial origin.

The anterior knee region is inseparably connected with the anterior femoral region. Within it, branches of the femoral artery and of its continuation, the popliteal artery, and of the anterior tibial artery unite to form a rich vascular plexus that nourishes the joint and the overlying skin. For details we refer to the anatomic literature where this region is described in considerable detail. Only the main organization will be sketched briefly.

Three main branches emerge from this plexus; two horizontal ones surround the anterior part of the knee joint, forming an arcade around it. They are crossed at a more or less acute angle by a third arcade which connects the inferior portion of the medial femoral surface with the tibiofibular joint.

The superior horizontal branch is formed by the medial and lateral superior geniculate arteries, which surround the knee in the form of an arcade at the level of the femoral condyles.

The inferior horizontal branch is created by the confluence of the medial inferior geniculate artery with the lateral inferior geniculate artery from the popliteal artery[165] at the level of the medial tibial condyle.

164. Sappey, p. 672.
165. Sappey, p. 680.

The supreme geniculate artery and the recurrent anterior tibial artery unite to form the third branch, which crosses the horizontal ones. In addition to these arteries, representing the principal branches of the vascular plexus of the knee, the recurrent posterior tibial artery and the superior fibular artery[166] also take part in the formation of the vascular plexus of the knee. Both of them are usually branches of the anterior tibial artery.

"All of these articular arteries anastomose in a finer, extensive network between the skin and the fascia, and in a stronger and narrower one which courses deep to the accessory ligaments of the joint and along the deep surface of the patella."[167] The superficial network spreads to the lower portion of the femoral region and to the upper portion of the lower leg.

A fairly extensive territory of the leg region (see Figures 16 and 17, Plate VIII) is associated with the popliteal artery. This is the territory which corresponds to the superficial part of the gastrocnemius muscle.

This muscle is nourished by the two sural arteries (medial and lateral), the largest branches of the popliteal artery.[168] They emerge from the main artery at the level of the knee joint, often from a common stem. The cutaneous branches of the popliteal artery that run to the inferior portion of the posterior femoral region can also emerge from them (author).

Both sural arteries descend to the corresponding head of the gastrocnemius muscle. Then they divide into one or more deep muscular branches, which spread out within the muscle, and one superficial branch (superficial sural artery, Krause).[169]

Overlying the gastrocnemius, the superficial branches travel inferiorly, parallel to the medial and lateral edge of this muscle. With their terminal ramifications, they extend as far as the skin overlying the Achilles tendon. According to Henle,[170] the lateral branch can often even be traced down to the ankle where it then anastomoses with the peroneal artery.

Both send out medial and lateral cutaneous branches at acute angles. These branches penetrate the thin sural fascia and spread out into the subcutaneous connective tissue of the sural skin territory.

In the furrow between the two heads of origin of the gastrocnemius muscle, a third artery emerges that participates in the supply of the skin of the sural area (observation of the author). One could call this the median superficial sural artery. It descends in close contact with the sural nerve (inferior saphenous nerve) deep to the fascia into the groove that the two gastrocnemius muscles form after their union. When it is strongly developed it can be traced beyond the middle of the leg. Medial and lateral branches arise from it at acute angles and perforate the fascia to course into the skin.

166. Henle, p. 199.
167. Henle, p. 207.
168. Henle, p. 679; Sappey, p. 196.
169. Krause, p. 1182; Theile, p. 233.
170. Henle, p. 196.

The origin of this artery is quite variable. Sometimes it emerges directly from the popliteal artery, sometimes from the lateral or medial sural artery, and sometimes from the common stem of the two.

Its expansion usually exists in distinct opposition to the appearance of these arteries. If the arteries are especially strongly developed, it appears as a minor secondary branch of one or the other. On the other hand, the lateral and medial superficial sural arteries often recede completely when the median sural artery is strongly developed. It is probably on this basis that Henle[171] describes only the lateral and medial superficial sural arteries in his discussion, whereas Sappey[172] mentions only the median superficial sural artery as a pronounced cutaneous branch of this territory.

However, the author of this work also came across cases where all three arteries were developed to approximately the same degree. Then, in the middle of the leg region, they blended into one another with numerous anastomoses.[173] According to the author's observations, the median superficial sural artery appears to be the most constant of the three.

In one case, for example (Figure 17, Plate VIII), the medial and lateral superficial sural arteries were only weakly developed. They were supplemented in the upper portion of the sural region by a number of small, direct cutaneous arteries from the popliteal artery and were replaced in the inferior portion by cutaneous branches from the deep muscle branches of the gastrocnemius muscle, which appeared in various positions. The median superficial sural artery was also present to the usual extent in this case.

The entire lateral leg cutaneous region, bounded anteriorly by the anterior ridge of tibia and posteriorly by the posterior edge of the fibula, is supplied by the anterior tibial artery. Of its musculocutaneous branches, the anterior and posterior recurrent tibial arteries and the superior fibular artery were already mentioned in connection with the arterial network of the knee joint, in the formation of which they participate. In its further course between the tibialis anterior, extensor digitorum, and extensor hallucis longus muscles, it gives rise to yet another large number of musculocutaneous branches. Henle mentions fifteen as the norm.[174] The cutaneous branches perforate the fascia in varying numbers at exit points which are not well specified (author). Usually they divide immediately after their emergence into ascending and descending branches that arborize in the subcutaneous fat.

The final territory of the leg region (see Figure 18, Plate VIII), which remained undescribed until now, is supplied by the posterior tibial artery. This territory is associated with the medial surface of the leg and is bounded posteriorly by the medial edge of the gastrocnemius muscle and anteriorly by the anterior border of the tibia.

171. Henle, p. 196.
172. Sappey, p. 679.
173. Figure 17 (Plate VIII) represents one such case.
174. Henle, p. 199.

The posterior tibial artery is the principal artery of this region. Along its entire course it sends out numerous musculocutaneous branches[175] that pass into the soleus, the flexor digitorum longus, and the tibialis posterior muscles, and into the overlying skin of the territory.

The larger cutaneous arteries, as a rule, appear from the fissure between the soleus and flexor digitorum longus muscles (author). Directly above the fascia, they divide into two diverging terminal branches, one passing anteriorly and one posteriorly. Both exhibit a highly tortuous course and a horizontal or weakly descending direction.

Numerous highly branched descending and ascending rami arise from them in a vertical direction and interconnect the terminal regions of the individual cutaneous arteries. In addition, they anastomose posteriorly with branches of the median superficial sural artery, into whose region they sometimes spread. Superiorly they anastomose with the superficial geniculate arterial rete, anteriorly around the anterior border of the tibia with cutaneous branches from the anterior tibial artery, and inferiorly with cutaneous arteries of the ankle joint.

For the region of the ankle joint and the foot we refer to the anatomic literature.

175. Arnold, p. 555.

Plates

Plate I

Figure 1

m_1-m_{12} Cutaneous twigs of the dorsal branches of the intercostal arteries.

m Cutaneous twigs of the dorsal branches of the lumbar arteries.

l Cutaneous twigs of the dorsal branches of the intercostal, lumbar, and sacral arteries.

pp Posterior perforating branches of the intercostal and lumbar arteries.

t Dorsal cutaneous branches of the transverse cervical artery.

ssa Superficial circumflex scapular artery.

dp Posterior subcutaneous deltoid artery.

ss Cutaneous branches of the supraspinatus fossa region.

ss^1 Cutaneous branches of the transverse scapular artery.

ss^2 Cutaneous branches of the transverse cervical artery.

c Cutaneous branches of the superficial cervical artery.

Figure 2

ts Superficial thoracic artery.

pl Lateral perforating branches of the intercostal and lumbar arteries.

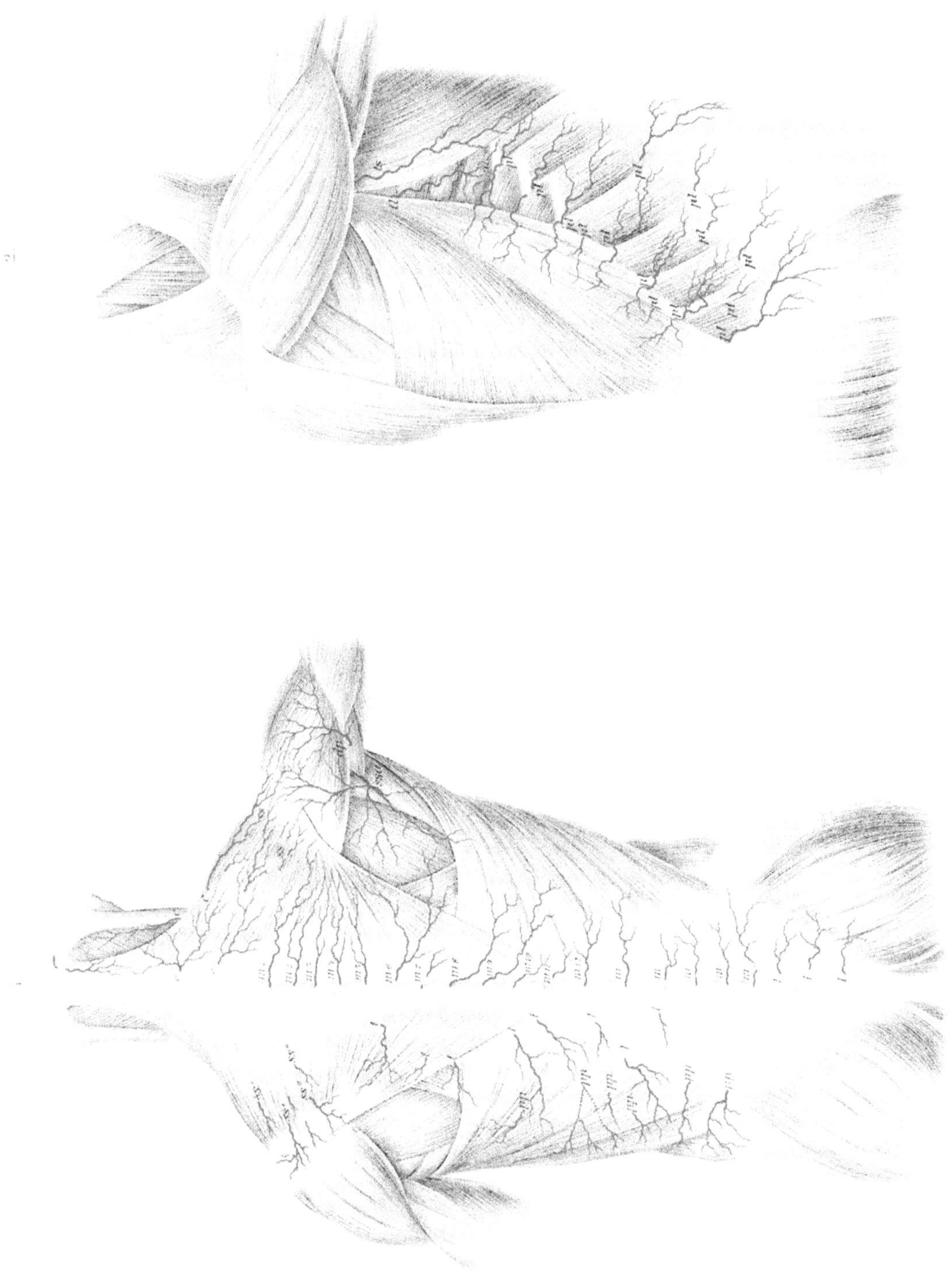

Plate II

Figure 3

ts	Superficial thoracic artery.
pa	Perforating branches of the anterior intercostal arteries.
pm	Perforating branches of the internal mammary artery.
pm_1	First perforating branch over the first rib.
pm_2	Second perforating branch (first intercostal space), etc.

Figure 4

tl	Long thoracic artery.
ts	Superficial thoracic artery.
ta	Thoracic branch of the thoracoacrominal artery.
pm	Perforating branches of the internal mammary arteries.
pm^1,pm^2	Double second perforating branch (first intercostal space).
pm^3	Third perforating branch (second intercostal space).
da	Subcutaneous anterior deltoid artery.
pa^1	Perforating branches of the anterior intercostal arteries.
pl	Lateral perforating branches of the intercostal and lumbar arteries.
ess	Superficial superior epigastric artery.
esi	Superficial inferior epigastric artery.
es	Cutaneous branch of the superior epigastric artery.
ei	Cutaneous branch of the inferior epigastric artery.
pra	Abdominal branch of the superior external pudendal artery.
cfs	Superficial circumflex iliac artery.

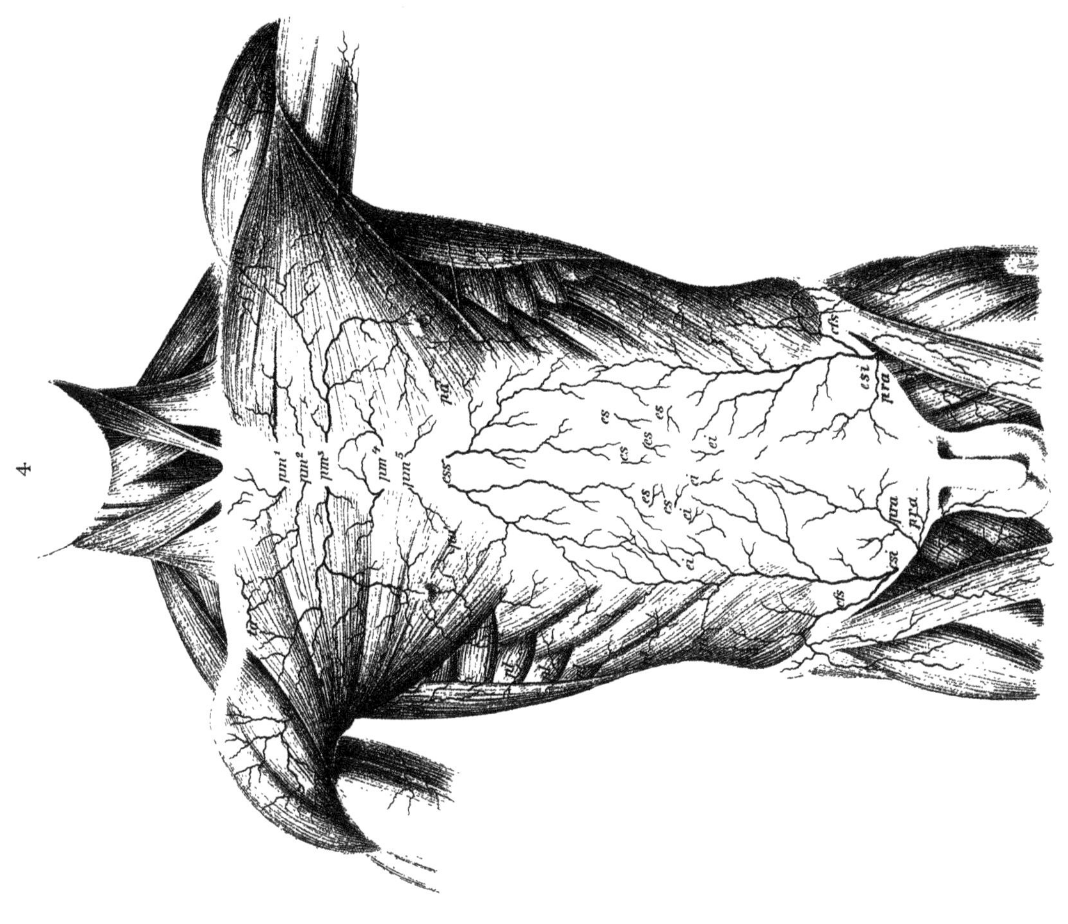

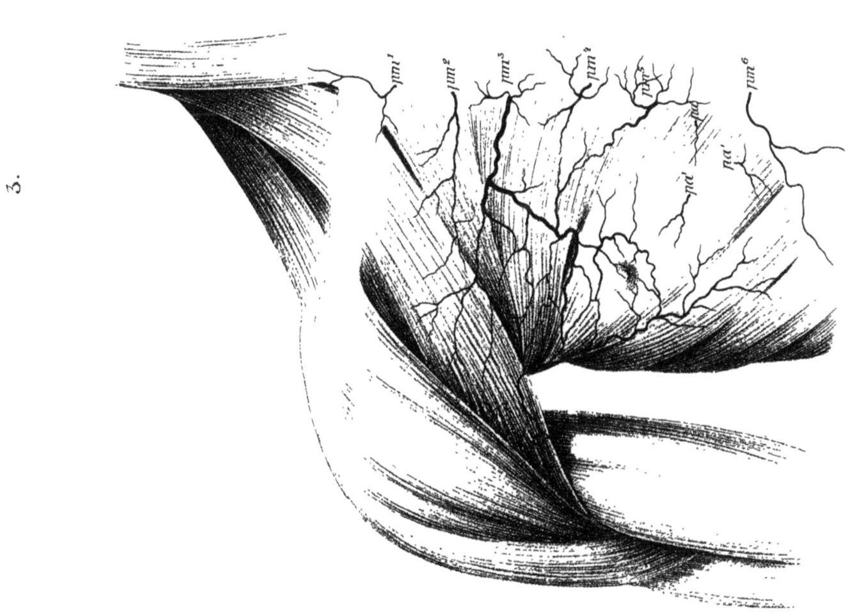

Plate III

Figure 5

pes Superior external pudendal artery.
   1 Abdominal branch.
   2 Pubic branch.
   3 Anterior scrotal branch.
   4 Cutaneous branches for the upper, medial thigh area.
pei Inferior external pudendal artery.
   5 Cutaneous branches for the upper, medial thigh area.
   6 Penile cutaneous branches.
   7 Anterior scrotal arteries.
dp Dorsal penile arteries.
 f Femoral artery.
fc Cutaneous branches of the femoral artery.
ce Common stem of the superficial inferior epigastric artery and the superficial circumflex iliac artery.

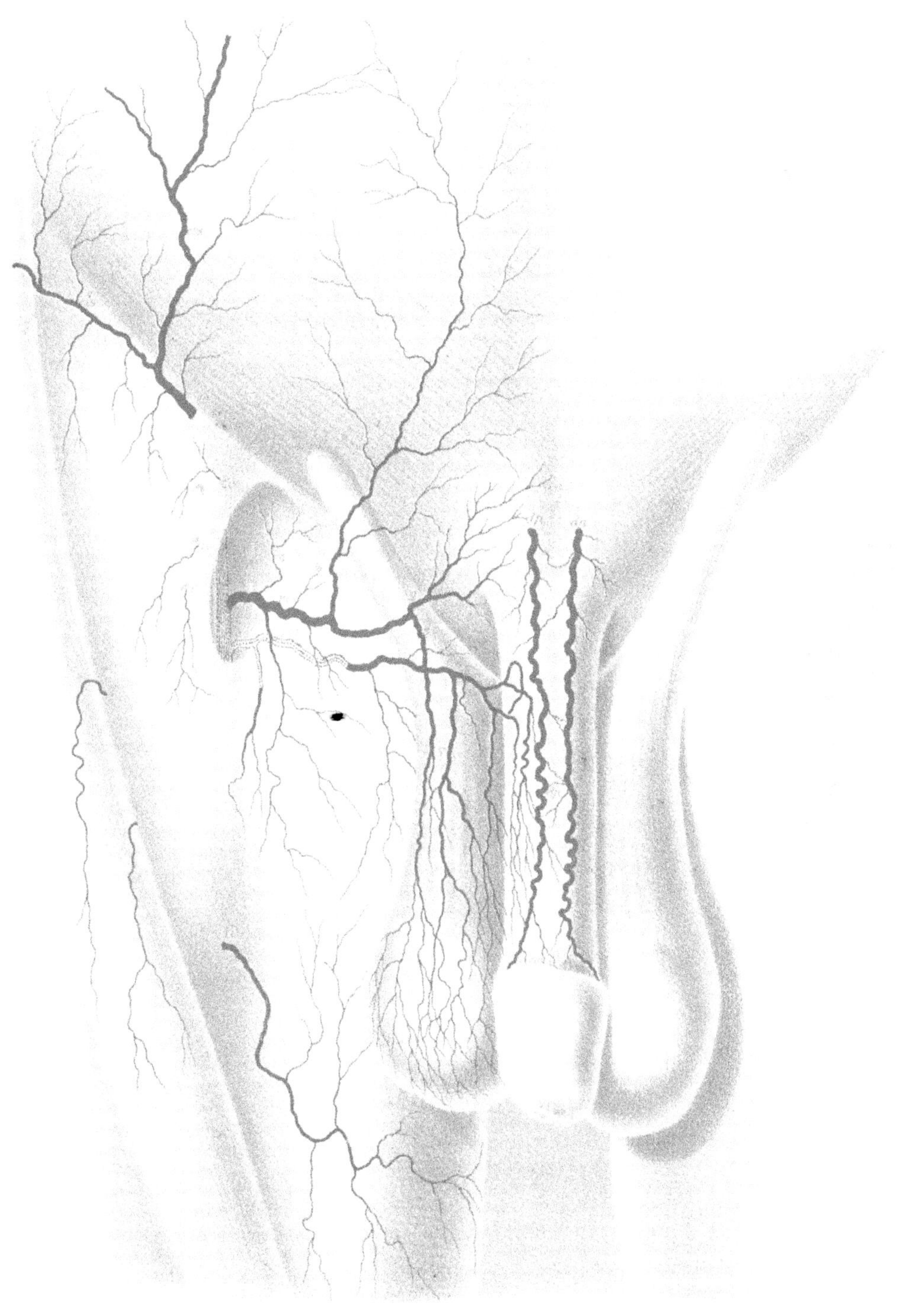

Plate IV

Figures 6 and 7

  1 Internal pudendal artery.
1a Gluteal cutaneous branches of the internal pudendal artery.
  2 External hemorrhoidal arteries.
2a Gluteal cutaneous branch of external hemorrhoidal arteries.
  3 Perineal artery.
3a Posterior scrotal arteries (posterior labial arteries in Figure 7).
3b Cutaneous branch of the upper medial thigh region.

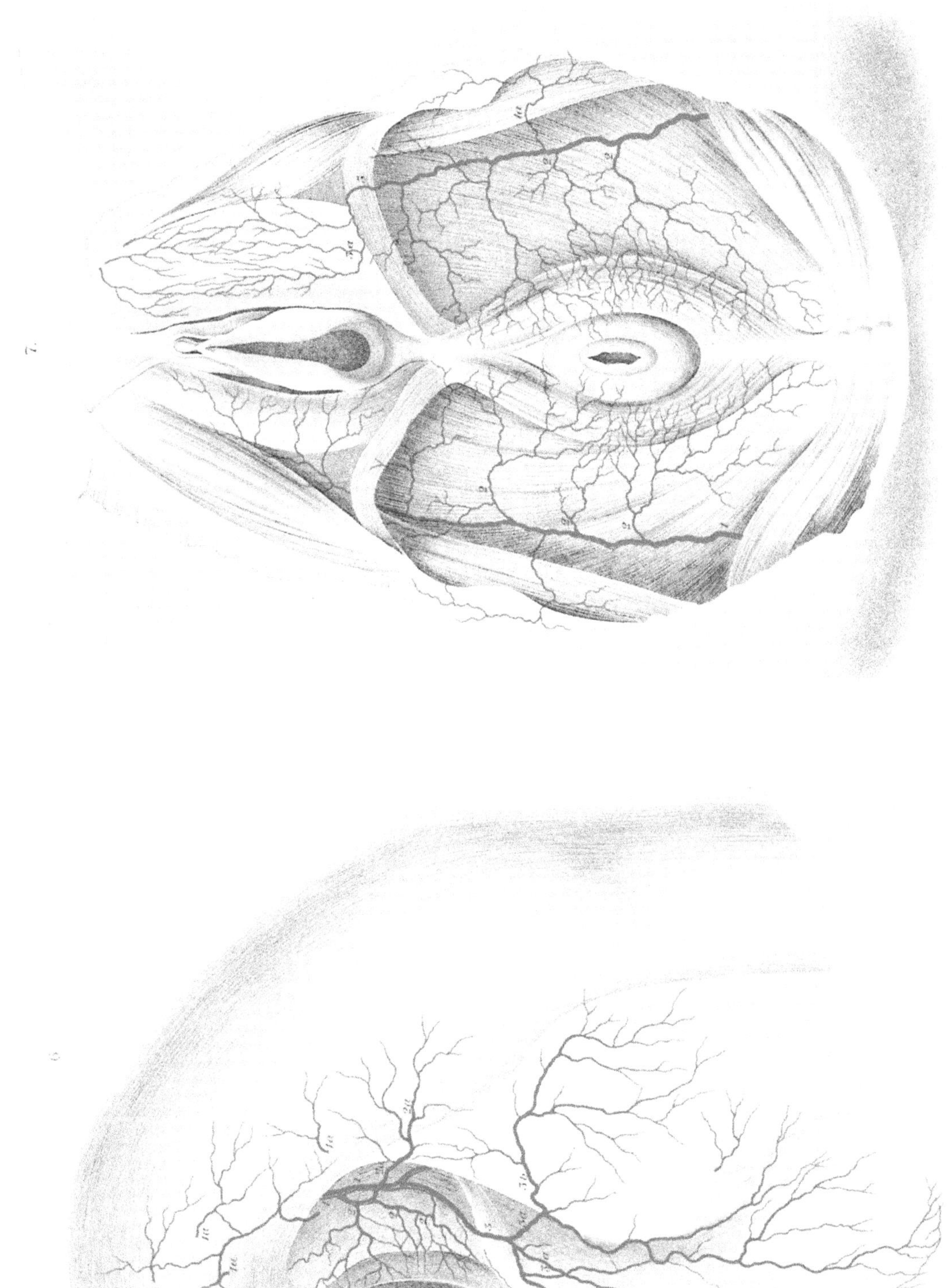

Plate V

Figure 8

    p   Subcutaneous parotid branches.
   sm   Cutaneous branches of the submental artery.
   cd   Descending cervical arteries from the occipital artery.
   cs   Cutaneous branches of the superficial cervical artery.
   tc   Cutaneous branches of the transverse cervical artery.
   ts   Cutaneous branches of the transverse scapular artery.
   ts¹   Subcutaneous supraclavicular artery.
        ra   Anterior branch.
        rp   Posterior branch.
        raa   Anterior auricular branches.
        rap   Posterior auricular branch.

Figure 9

    1   Posterior auricular artery.
   2,3   Posterior auricular arteries.
    4   Temporal artery.
    5   Anterior auricular artery.

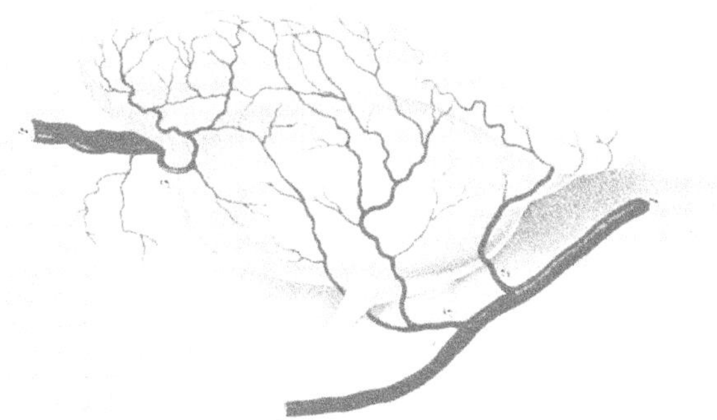

Plate VI

Figure 10

    1 Subcutaneous posterior deltoid artery.
    2 Inferior radial collateral artery with its cutaneous branches.
    3,4 Cutaneous arteries from the brachial artery.

Figure 11

    1 Subcutaneous anterior deltoid artery.
    2 Brachial artery.
    3 Superficial branch of the superior ulnar collateral artery with its cutaneous branches.
    4–15 Cutaneous branches of the brachial artery.
    16 Cutaneous branch of the profunda brachii artery.

Figure 12

    1 Brachial artery.
$\alpha,\beta,\gamma,\delta$ Cutaneous arteries from the brachial artery.
    2,3,4 Cutaneous branches of the median artery.
    5 Superficial antecubital fold artery from the brachial artery.
    6 Radial artery.
    7 Cutaneous branches of the recurrent radial artery.
    7^1 Cutaneous branches of the radial artery.
    8 Cutaneous branches of the ulnar artery.

Figure 13

    1 Recurrent interosseous artery.
    2–12 Cutaneous branches of the posterior interosseous artery.
    13,14 Cutaneous branches of the ulnar artery.
    15–17 Cutaneous branches of the anterior interosseous artery.
    18 Radial collateral artery.

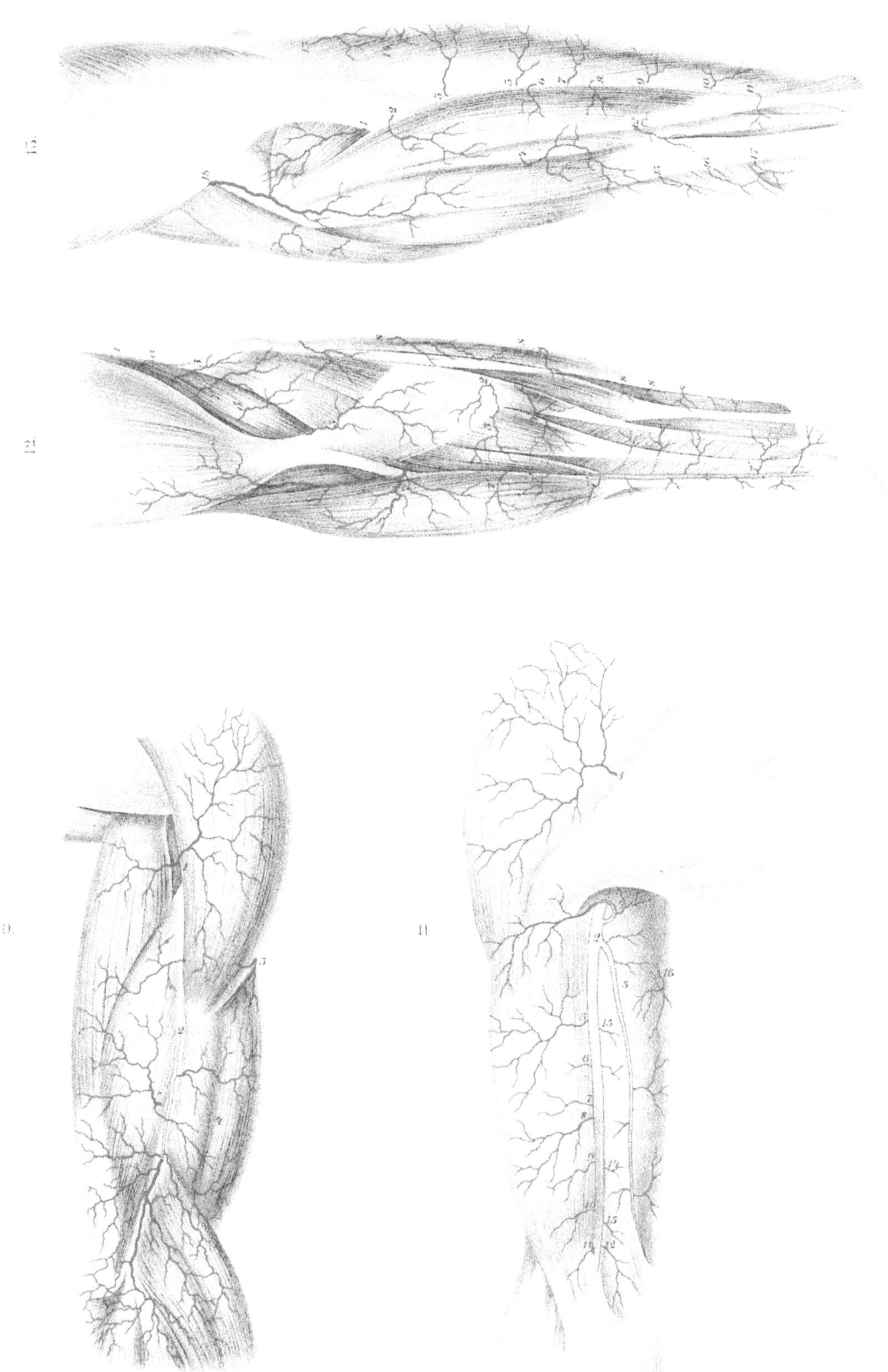

Plate VII

Figure 14 (The upper portion of the sartorius muscle has been removed.)
1 Femoral artery.
α Cutaneous branches of the femoral artery.
2 Profunda femoris artery.
β Cutaneous branches of the profunda femoris artery.
3 Superficial geniculate arterial rete.

Figure 15
1 Cutaneous branches of the superior gluteal artery.
2 Cutaneous branches of the inferior gluteal artery.
3 Cutaneous branches of the internal pudendal artery.
4 Cutaneous branches of the lateral sacral arteries.
5 Cutaneous branches of the iliolumbar artery and of the last lumbar artery.
6 Cutaneous branches of the obturator artery.
7 Cutaneous branches of the medial circumflex femoral artery.
8 Cutaneous branches of the perforating arteries from the profunda femoris artery.
9 Cutaneous branches of the popliteal artery.
10 Cutaneous branches of the lateral femoral circumflex artery.

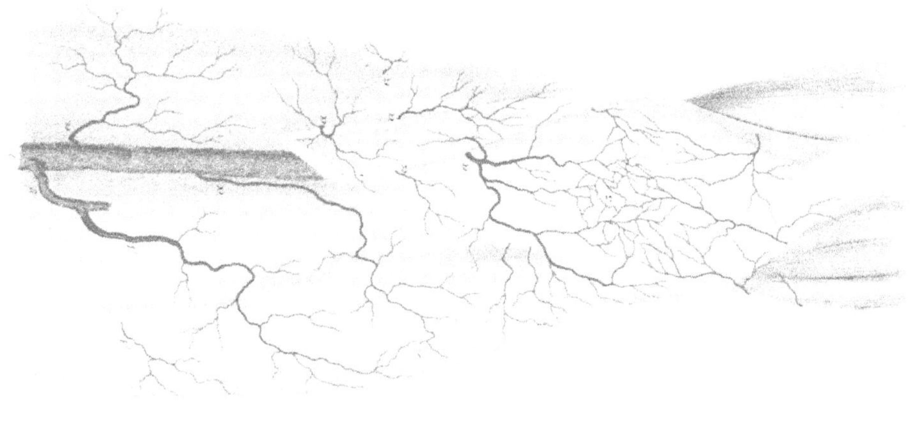

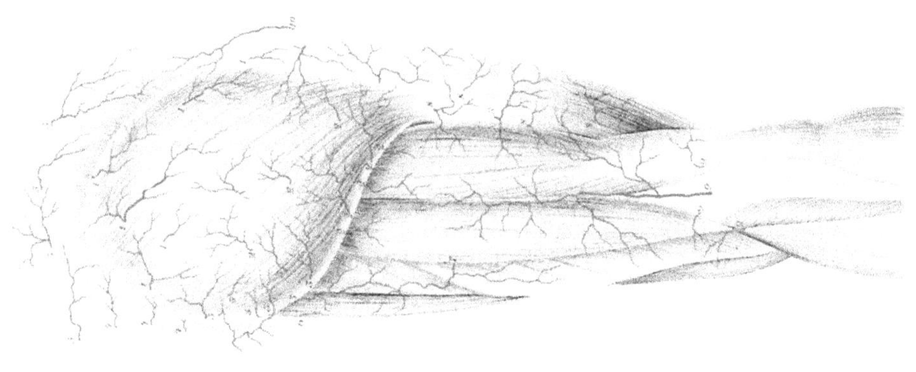

Plate VIII

Figures 16 and 17
1 Superficial medial sural artery.
2 Superficial lateral sural artery.
3 Superficial median sural artery.
4 Musculocutaneous branches of the deep sural arteries.

Figure 18 Cutaneous branches of the posterior tibial artery at the inferomedial thigh area.

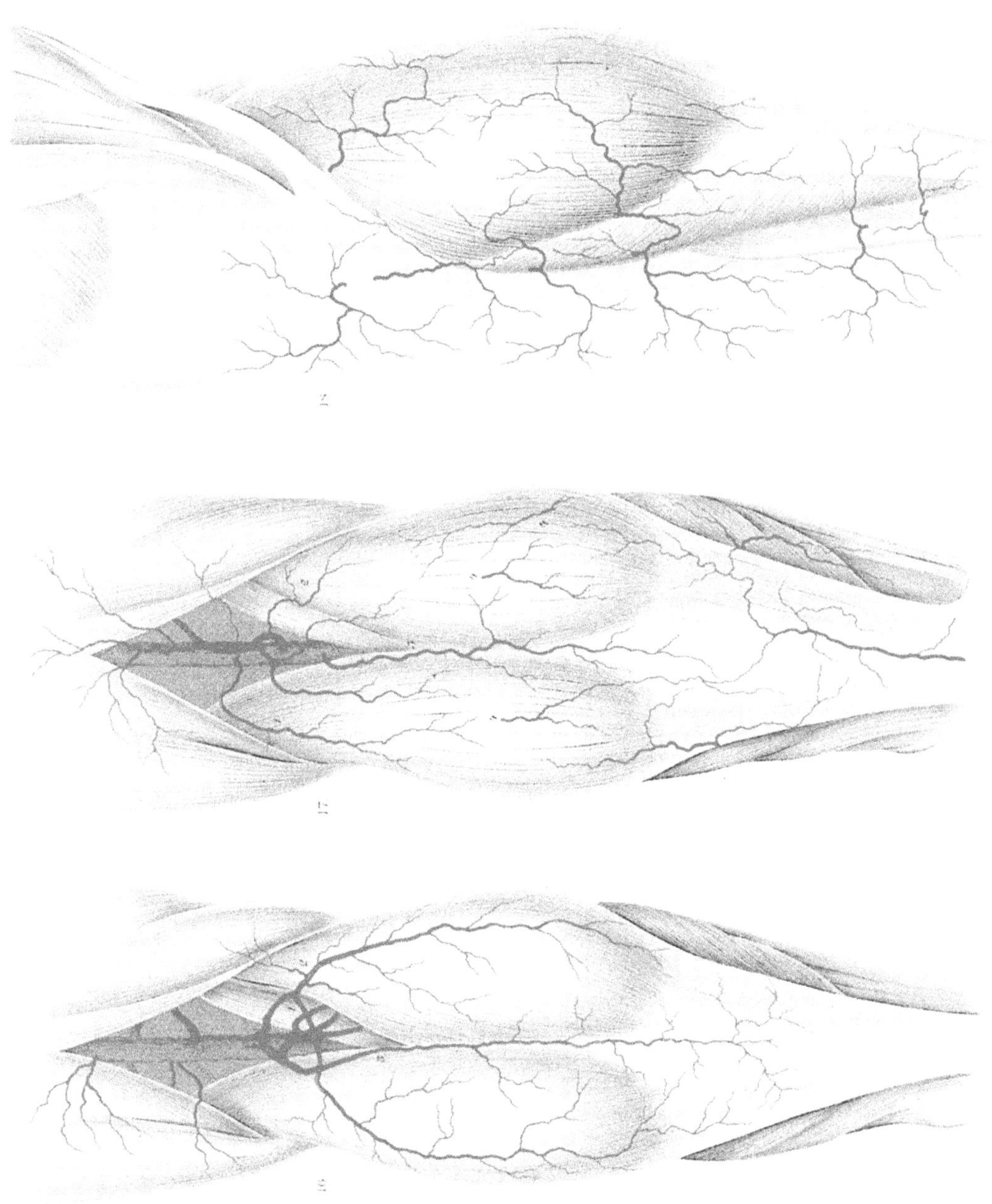

Plate IX
Overview of the Cutaneous Arterial Territories

Figure A

1 Cutaneous territory of the superficial superior epigastric artery.
2 Cutaneous territory of the superficial inferior epigastric artery.
3 Cutaneous territory of the superior and inferior epigastric arteries.
4 Cutaneous territory of the external pudendal arteries.
5 Cutaneous territory of the dorsal penile arteries.
6 Cutaneous territory of the perforating branches from the intercostal arteries.
7 Cutaneous territory of the perforating branches from the lumbar arteries.
8 Cutaneous territory of the superficial circumflex iliac artery.
9 Cutaneous territory of the profunda femoris artery (circumflex femoral arteries).
10 Cutaneous territory of the femoral artery.
11 Cutaneous territory of the superficial genicular rete.
12 Cutaneous territory of the anterior tibial artery.
13 Cutaneous territory of the posterior tibial artery.
14 Cutaneous territory of the popliteal artery (sural arteries).
15 Cutaneous territory of the thoracic arteries.
15a Cutaneous territory of the thoracoacromial artery.
16 Cutaneous territory of the perforating branches of the internal mammary artery.
17 Cutaneous territory of the thyrocervical trunk.
18 Cutaneous territory of the superior thyroid artery.
19 Cutaneous territory of the subcutaneous anterior deltoid artery.
20 Cutaneous territory of the brachial artery.
21 Cutaneous territory of the superior ulnar collateral artery.
22 Cutaneous territory of the radial artery.
23 Cutaneous territory of the median artery.
24 Cutaneous territory of the ulnar artery.

Figure B

1 Cutaneous territory of the dorsal branches from intercostal arteries.
2 Cutaneous territory of the dorsal branches of the lumbar arteries.
3 Cutaneous territory of the dorsal branches from the sacral arteries.
4 Cutaneous territory of the posterior perforating branches of the intercostal arteries.
5 Cutaneous territory of the posterior perforating branches of the lumbar arteries.
6 Cutaneous territory of the thyrocervical trunk.
   a Of the superficial cervical artery.
   b Of the transverse scapular artery.
   c Of the transverse cervical artery.
7 Cutaneous territory of the subcutaneous posterior deltoid artery.
8 Cutaneous territory of the superficial circumflex scapular artery.
9 Cutaneous territory of the inferior radial collateral artery.
10 Cutaneous territory of the superior ulnar collateral artery.
11 Cutaneous territory of the cubital rete.
12 Cutaneous territory of the radial artery.
13 Cutaneous territory of the ulnar artery.
14 Cutaneous territory of the external and internal interosseous artery.
15 Cutaneous territory of the superior gluteal artery.
16 Cutaneous territory of the inferior gluteal artery.
17 Cutaneous territory of the internal pudendal artery.
18 Cutaneous territory of the obturator artery.
19 Cutaneous territory of the perforating branches of the profunda femoris artery.
20 Cutaneous territory of the popliteal artery.
21 Cutaneous territory of the anterior and posterior tibial artery.

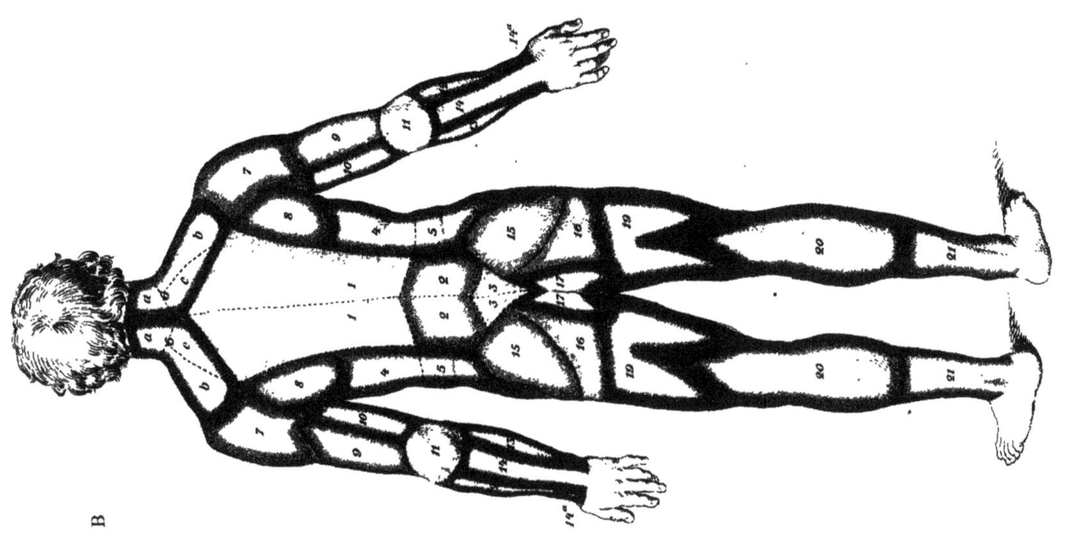

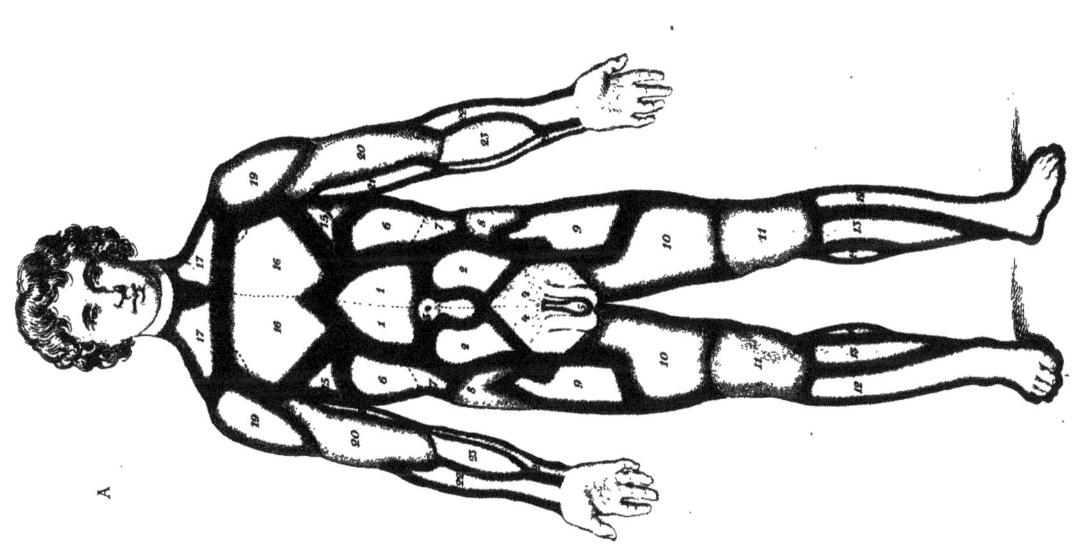

Second Part

From the description of the cutaneous arteries, it becomes apparent that their characteristic features are identified by those details that were until now noticed the least. Their origins from the deep arteries are subject to wide variations in some areas. The exit points from the muscular layer of the body cannot always be specified precisely and in a generally accepted manner. However, they demonstrate great regularity in their regions of arborization, especially in their arrangement and orientation. This relatively greater regularity in regions of arborization compared with their origins is a feature they share in common with all other arteries of the body. The terminal ramifications and the territories supplied are always much more constant than the origin and the stem arteries from which they arise. It is sufficient to refer to the cutaneous arteries of the neck or of the lateral chest wall as examples. There the peripheral region of expansion and distribution often provides the only truly distinctive characteristic. This phenomenon is based on the development of the vascular system.

The usual description of the arteriovascular system rests on the image of a tree, whose root is the heart and from whose trunk branches sprout to the parts of the body. Ontogeny, on the other hand, teaches that the primitive vessels grow into the embryonic organism from its periphery. At least for reptiles, birds, and mammals, this has been proven by the well known investigation of His, which found further supplementation and verification in the works of Türstig. The terminal ramifications of the customary representation are thus in reality the earlier, original ones. Only secondarily do they flow together to form the large trunks.

Still more consistent and more characteristic of the cutaneous arteries are their relationships in terms of arrangement and orientation. These are all the more remarkable since, with the cutaneous arteries, one least expects to find extensive regularity in this area. This must have its basis in certain factors.

The fact that the ramifications of the arteries developed before the trunk makes it very probable from the start that the basic metameric layout of the arterial system, as it is found in the stronger vessels of the body, also comes into play in the arrangement of the cutaneous arteries as well. This basic metameric layout of the vascular system is a mechanical consequence of the organization of the embryo into a number of individual, sequentially aligned pieces, originally of equivalent value.

The primitive vessels form a net in the area vasculosa, from which solid shoots grow into the embryo. The direction of their advance is therefore determined by the least resistance. In this fashion, the basic metameric organization of the body is expressed also in the vascular system.

The results described above (First Part) of a detailed examination of the cutaneous arteries of the body present an entire group of facts which can be traced back directly to the metamerism of the body. The cutaneous branches of the dorsal rami from the intercostal and lumbar arteries, the cutaneous branches of the sacral region (from the lateral sacral arteries), the direct perforating cutaneous branches from the intercostal and lumbar arteries in the lateral thoracic and abdominal regions, all emerge from the muscular layer, each between two thoracic vertebral segments, and arborize in the skin parallel to the ribs, generally in an arrangement corresponding to the vertebral segments.

Also, in the anterior chest region, the metameric organization can still be partially recognized. However, as already pointed out, the variations here emerge in which the arrangement no longer corresponds exactly to the vertebral segments.

Besides those numerous variations occurring in the vascular system that may sometimes be brought about by uneven development and expansion in the customary behavior of arteries of equivalent strength, there are certain regions of the skin in which the arrangement corresponding to the vertebral segments is regularly modified. These include the portion of the back corresponding to the trapezius muscle and the anterior thoracoabdominal region.

In the former, the five or six superior cutaneous arteries converge on the triangular origin of the scapular spine, whereas in the latter, the image of the basic metameric organization is largely obscured by the considerable development of the perforating branch (from the internal mammary artery) that emerges from the second intercostal space.

The metamerism of the cutaneous arteries, still evident from the organization, is modified and disguised in this region as a result of the influence of a second factor. This other factor, also affecting the direction and arrangement of the cutaneous arteries, has to do with the relationship of the covering integument to body growth.

In attempting to clarify the influence of skin and body growth on the arrangement of the cutaneous arteries, a gap in our knowledge of the development of the vascular system is very much in evidence.

The primordial vascular organization and that of the adult are well known. Extremely little is known about the stages of growth lying between. Above all, there is no precise information detailing to what extent the development of the vascular system consists only in an increase in size and outward growth of the vessels from the area vasculosa, and to what extent this may be influenced by the secondary sprouting of new arteries from the first vessels. Until there is a more exact answer to this question, many things will remain uncertain in ascribing the influence of growth relationships on the arrangement of the cutaneous arteries.

If one assumes, along with Sappey and others,[176] that the area vasculosa is of preeminent importance in the formation of the vascular system, then it is clear that the cutaneous arteries must follow the extent of the expansion or the migration of the cutaneous regions that they supply, for they are fixed in these cutaneous regions with their terminal ramifications.

In its growth and extension, the skin is absolutely dependent on the growth of the body it encloses. It must follow every curve or convexity that the body experiences in its growth, and it must adapt to every new appendage by protruding and enclosing it. It follows from this, as was noted above, that the growth processes of the body must also exert influence on the arrangement of the cutaneous arteries.

In actual fact, it appears that many peculiarities in the arrangement of the cutaneous arteries may be attributed to the growth processes of the body, either directly or indirectly.

The basis for some interrelationships in the arrangement must be researched in the growth of the embryo. In the abdominal region, the superficial superior and inferior epigastric arteries, as well as the abdominal branches of the external pudendal artery, are noteworthy as a result of the contrast in which they stand in relationship to the metamerism of the remaining cutaneous arteries of the body. They cross the paths of the standard cutaneous arteries at a right angle and arborize in the region of various metameres.

If one now observes an embryo, in which the body is sealed off ventrally except for the navel, the strong curvature of the entire trunk is its most conspicuous feature. The forehead and caudal end are almost touching. The anterior abdominal wall is very poorly developed. The exit points of the superficial superior epigastric artery on the one side and the superficial inferior epigastric artery and external pudendal arteries on the other side, are actually only separated from one another by the umbilical cord. Later, the marked curvature of the embryo disappears with extension of the spine. The necessary consequence of this process is the expansion of the anterior abdominal region. The exit points of these arteries just mentioned move further and further apart. In a fully developed condition, both are far removed from the navel. The terminal ramifications of the cutaneous arteries must follow the expansion of the territories which

176. Sappey, p. 522.

they supply. Since this is primarily an expansion in the longitudinal direction, but also somewhat in width, the great sheet-like expansion of these arteries with predominant vertical orientation clearly results from this, as described in the first chapter of this treatise.

Perhaps the embryonic history may also give some explanation about the causes for the arrangement of the cutaneous arteries in the neck and throat region.

The cutaneous arteries of the neck emerge from the muscular branches of the trapezius muscle. These in turn emerge for the most part from the arteries of the lateral neck region (superficial cervical artery, transverse cervical artery, and transverse scapular artery; see Chapter IV).

The explanation for this arrangement is found in the development of the superficial cervical layers. According to His,[177] it is very probable that the neck portion of the trapezius muscle does not belong to the true neck muscles but is rather diverted secondarily into the neck region from the side of the neck.

A second surprising manifestation of the cutaneous arteries of the neck is the fact that no metamerism appears detectable in the organization at first glance. All of these exhibit a diagonally lateralward descending main direction. This condition can perhaps also be explained through the developmental history of the neck. In the youngest embryos, the primitive vertebrae of the neck portion have a surprisingly deep location when compared with the thoracoabdominal organs. His[178] states that in their case, the lower boundary of the primitive cervical spine "reaches down below the level of the hepatic duct and into that of the umbilical artery. In somewhat older embryos as well, the primordia of the liver, stomach, lungs, and heart are still in the region of the cervical primitive spine. This condition only disappears gradually by means of the combined efforts of the two factors.

While the cupola of the coelomic cavity descends, the rudimentary cervical spine ascends until its lower end reaches the cupola of the coelomic cavity. It is not our task to describe these processes in detail. It is only of importance here that consequently ". . . with progressive development, the corresponding metameric zone experiences an increasing inclination, which also finds lasting expression in, among other things, the orientation of the fibers of the lateral neck musculature (with the inclusion of the trapezius muscle)." On this basis, it would perhaps be possible to regard the diagonally descending arrangement of the cutaneous arteries of the neck parallel to the fibers of the trapezius muscle as a similar modification of the metameric basic structure. The reason for this modification could then be sought in the upward movement of the rudimentary cervical spine.

Apparently the manner of development of the anterior neck and submandibular region, bounded posteriorly by the two sternocleidomastoid muscles, helps

177. His, Anatomie menschlicher Embryonen III, p. 114.
178. His, Anatomie menschlicher Embryonen, p. 117.

to explain the arrangement of the cutaneous arteries. It is not important whether we decide on His' or Rabl's conceptualization of the development of the neck.[179] The descriptions of both authors are based on the fact that in the embryo during a specified period, a true neck, i.e., a coelom-free portion of the body interposed between chest and head, is not formed. The Meckel's cartilage, from which mandibular development proceeds, reaches inferiorly to the primary thoracic cavity.

The end result of the developmental processes that bring about the formation of the neck is the distance of the mandible from the sternal notch, which the skin must follow. Thus the anterior cervical region develops in similar fashion to the anterior abdominal surface. Since the expansion of the anterior neck skin follows a predominantly longitudinal direction, the cutaneous arteries of this region exhibit a vertical (ascending or descending) course in their main orientation.

If this interpretation is correct, it then also becomes clear why the superior portion of the cutaneous region, the submandibular triangle, which extends to the inferior edge of the larynx, is associated with the submandibular portion of the head according to its cutaneous arterial origins, while the cutaneous branches of the inferior neck portion ascend from the superior edge of the thoracic region.

When one considers the arrangement of the cutaneous arteries of the face, some causative factors must be sought in the embryonic growth processes as, for example, in the cutaneous arteries of the nose.

In addition, this appears to be very probable for the external ear. His has shown that the external ear develops from the hillocks surrounding the first branchial cleft. "Early on, their arrangement already shows a segmentation into a row of more or less sharply pronounced protuberances. In embryos from the end of the first month, six of these can be differentiated, of which two are associated with the mandibular arch (one and two), three are associated with the second pharyngeal arch (four, five and six), and one is associated with the combining piece between the first and second pharyngeal arches"[180] (three). The individual parts of the external ear arise from them.[181] The tragus of the developed ear arises from tuberculum one of His or the tragicum, tubercula two and three unite in the formation of the helix, tuberculum four becomes the anthelix, tuberculum five becomes the antitragus, and the earlobe emerges from tuberculum six.

It is thus very interesting to see that the arteries from the superficial temporal artery supply only those parts of the ear that arose from the anterior edge of the branchial cleft (tubercula one and two), i.e., the tragus and the

179. Compare His, Ueber den Sinus praecervicalis und ueber die Thymusanlage. (Archiv fuer Anatomie und Physiologie). His, Die Retromandibularbucht. (Anatom. Anzeiger 1886, Nr. 1). Rabl, C., Zur Bildungsgeschichte des Halses. (Prager medizinische Wochenschrift 1886, Nr. 52). Rabl, Ueber das Gebiet des Nervus facialis. (Anatom. Anzeiger 1887, Nr. 8).

180. His, Anatomie menschlicher Embryonen III. pp. 211–221. Leipzig, F.C.W. Vogel.

181. Compare the figures in His.

anterior portion of the helix. On the other hand, those structures arising from the posterior edge of the branchial cleft (tubercula four, five, and six, i.e., anthelix, antitragus, and earlobe) are associated with branches from the posterior auricular artery. In the region of tuberculum three, bow-shaped anastomoses between the branches of both groups finally occur. This curious arrangement is undoubtedly explained by the developmental history of the external ear. On the other hand, the accuracy of His' description receives new corroboration from it.

The embryonic processes of growth and development are undoubtedly of great importance as well in the arrangement of the cutaneous arteries of the external genitalia. The exit of the testes from the visceral cavity and the resulting protrusion of the scrotum, the development of the large labia of the vulva, the penis, etc., have found permanent expression in the orientation and arrangement of their cutaneous arteries.

Finally, the rudimentary budding of the extremities will also exert a direct influence on their own cutaneous arteries. In addition, the germination of new arteries from the original trunk appears to be precisely significant for the development of the arterial vascular system of the extremities, according to the opinion of the author. The arteries of the extremities are thereby placed with particular contrast to the arteries of the trunk. Were the growth of the vascular system of the extremities predominantly an enlargement of the rudimentary layer originally formed out of the area vasculosa, then it should be possible, in keeping with our previous statements, to trace homogeneous cutaneous arteries from the base of the skin protrusion of the extremity to its tip. These cutaneous arteries would then be, so to speak, taken along and stretched out in length in proportion to the growth of the extremity. This should result in an arrangement of the cutaneous arteries similar to that which in fact exists in the skin protrusion of the penis, the scrotum, and the labium majora. This is, however, by no means the case. On the contrary, the skin of the extremities (see Chapters IV and V of the treatise) is supplied by a large number of small arteries that arise from the large vessels of the extremities along their entire course. But without a doubt, the growth and stretching alterations of the skin are also of decided importance to the orientation of these secondarily developed cutaneous vessels.

The effect of the growth relationships of the body on the skin and its growth has received new illumination through the investigations of Langer[182] "on the cleavability of the skin" and "on the stretchability of the skin."

Dupuytren had made the observation that the puncture of a round instrument, e.g., an awl, does not produce any round holes, but rather linear clefts. At the same time, he noticed that the direction of these clefts is different in different parts of the body. This observation was utilized by Langer in methodical experiments. He thus obtained systems of lines that "in the manner of tattooing,

---

182. Langer, Zur Anatomie und Physiologie der Haut.I.Ueber die Spaltbarkeit der Cutis. II. Die Spannung der Cutis. Sitzungsbericht der mathem.-naturw. Classe der Wiener Akademie der Wissenschaften. XLIV. Bd. I Abth.

outlines the body with great regularity of form and "within certain limits, this was repeated in all individuals examined." Langer now shows that the direction of cleavability of the skin is directly related to its tension. A continuous stretching of the skin in a particular direction has as a consequence the shifting of the fibrillar orientation of the skin in that direction. The tension relationships, in turn, are based on the circumstances of growth of the entire body, especially in the extremities.

This interpretation, already self-evident for physical and mechanical reasons, is also confirmed by observation on the stretching of the abdominal skin in pregnant women (postpartum stretch marks) and the extensibility of skin associated with tumors, etc.

Therefore, if the conjecture that the arrangement and orientation of the cutaneous arteries are influenced significantly by the processes of growth is correct, then agreement must exist, within certain limits, between the orientation of the lines of cleavage and the orientation of the cutaneous arteries.

A comparison of the figures that accompany this work with the plates from Langer's treatise in fact results in a surprising confirmation. One thing must be borne in mind in making this comparison, however. The circumstances of embryonic growth, whose influence on the orientation of certain cutaneous arterial trunks was discussed above, will naturally also be expressed in these lines of cleavage. But the influence of the embryonic growth conditions may be obscured because of later growth alterations. For this reason, a dissimilarity in the main direction of these cutaneous arterial trunks and the orientation of the lines of cleavage occurs in some places, e.g., in the anterior abdominal and anterior cervical regions, in the superolateral thorax, and so forth. Langer has in fact made the observation that in embryos, and even in newborns, the cleavage lines differ substantially from those which he found in adults. He is also of the opinion[183] "that development and growth conditions are expressed in this diversity." Unfortunately, he did not publish his observations on the cleavability of the skin in embryos and newborns.

Without doubt, the degree of agreement among cutaneous arteries of embryos and newborns would be found to be greater with those arteries whose orientation is primarily dictated by embryonic growth circumstances. Taking this condition into account, one will find no territory wherein a fundamental difference could exist between the orientation of the cutaneous arteries and the lines of cleavage.

Extensive agreement is evident over all cutaneous regions of the head, neck, back, genitalia, thorax, and lateral abdominal wall. However, this holds true in a very special way for the cutaneous arteries of the extremities.

It was already pointed out that the arrangement of the cleavage lines is also a function of postembryonic growth processes, and that these can even obscure the influence of the embryonic growth patterns. These postembryonic

---

183. Langer, p. 24.

growth processes thereby gain significance for the arrangement and orientation of the cutaneous arteries.

This postembryonic development is of greater consequence for the extremities and the direction of stretching of their skin than for the trunk. According to Langer,[184] the mechanisms of stretching of the extremities, as found in the adult, are largely acquired later, and differ from the original situation. According to their original arrangement, the lines of cleavage are arranged across the extremities, circling their contents. Later, they alter to a longitudinal course. This shifting of the direction of cleavage lines leads Langer especially to consider the functional use of the extremities. "The joints, which are flexed at an angle during intrauterine life, are gradually brought to their full extended position after birth; the skin is thereby stretched in the longitudinal direction of the limbs, the fibrous network is stressed in this direction and held permanently by means of growth of the extremities, which occurs rapidly during this period of life."[185] Perhaps Langer overestimates the importance of the joint extension when compared with the importance of overall growth of the extremities. The longitudinal growth is so significant that the longitudinal reorientation of the cleavage lines could undoubtedly also be ascribed to it.[186]

Besides that, the extension of the joints can contribute to longitudinal reorientation only on the flexor side of the extremity. In fact, on the extensor side, the longitudinal extension that is brought about by longitudinal growth is partially cancelled out again through the stretching of the joints. This probably accounts for the arterial networks on the extensor side of the knee and elbow joints, in which no particular direction predominates, and for the horizontal orientation of the cutaneous arteries on the extensor sides of the extremities.

———————•———————

In the previous section an attempt was made to trace the arrangement of the cutaneous arteries to the fundamental metameric organization of the vascular system and to the growth relationships of the body and the skin.

If this representation is correct, then those features that are governed by similar conditions of arrangement must exhibit some similarity to the arrangement of the cutaneous arteries.

Without a doubt, the arrangement of the nerves, veins, and lymphatic vessels of the skin is related to the cutaneous arteries in a particular way. Indeed it even appears that the orientation of the hair in some regions exhibits conformity with the orientation of the cutaneous arteries. Although it appears strange, this

184. Langer, p. 151.
185. Langer, p. 152.
186. The lower extremity, for example, as compared with the total body length, is in a ratio of 1:2.5 for a newborn, and a ratio of 1:1.8 for an adult.

becomes understandable when one considers that the orientation of both is to a great extent determined by the growth circumstances (ability to split) of the skin.[187] It would certainly be interesting to carry out a comparison for all of these cutaneous organ systems. At this point, only the cutaneous nerves will be discussed.

The form and orientation of the expansion of the cutaneous nerves and the extent of their territories depends primarily on the fundamental metameric organization of the peripheral nervous system and secondarily on the development and growth circumstances of the body and skin. In comparing the cutaneous arteries with the cutaneous nerves, however, one must not forget that structures as differently organized as our nerves and arteries often adapt in completely different ways to the same conditions. The simple fact that the arteries develop primarily from the periphery to the center, whereas the nerves grow out from the central structure to the periphery, will have considerable significance.

In fact, it appears that the nerves are governed by completely different rules of growth than the arteries. Therefore, the growth circumstances of the body will also often be expressed in a completely different manner in the nerves.

A detailed comparison discloses the fact that the degree of correspondence between cutaneous arteries and cutaneous nerves is greater in those regions where the influence of the fundamental metameric organization is more in evidence.

The more the growth conditions determine the organization of the arteries and nerves of a territory, the smaller is the correspondence between the two.

In the neck region,[188] for example, a comparison between cutaneous arteries and cutaneous nerves leads to negligible correspondence. The fundamentally different arrangement of the large nerve and arterial trunks precludes any comparison of their origins from the outset. In the regions of terminal expansion, as well, and even in the directional relationships, no clear pattern of similarity emerges whatever. The same holds true for the cutaneous territories of the extremities. An attempt at reconciling the regions of expansion of the cutaneous nerves and cutaneous arteries of the extremities leads to artifacts, but not to any positive results. Nowhere does any very obvious correspondence occur. Only in the orientation of the cutaneous arteries and cutaneous nerves does a more or less pronounced similarity exist in some territories.[189]

In contrast, in the skin of the trunk, in which the fundamental metameric organization is most clearly in evidence, the correspondence is surprising. In

187. Compare Voigt, Die Richtung der Haare am menschlichen Koerper. Langer, Ueber die Spaltbarkeit der Cutis. p. 44.

188. The information about the distribution of the cutaneous nerves is taken from the works of Voigt, Sappey, Henle, Schwalbe, and Heiberg. Also compare Plate IX of this treatise.

189. Thus for example in the gluteal region, on the posterior surface of the upper thigh, on the knee, on the calves, the shoulders, etc.

the first chapter of this treatise, the boundaries of the cutaneous territory supplied by the dorsal rami of the intercostal and lumbar arteries as well as the cutaneous branches of the lateral sacral arteries were described in detail. This region of peripheral expansion corresponds for the most part with that of the cutaneous branches from the dorsal rami of the thoracic, lumbar, and sacral nerves. In only two areas do the latter extend significantly beyond this territory.

Whereas the cutaneous branches of the upper intercostal arteries reach only to the triangular origin of the spine of the scapula with their terminal ramifications, the analogous cutaneous nerves advance as far as the acromion and also supply the skin overlying the subscapular fossa.[190] Further, in the hip region the dorsal cutaneous branches of the lumbar and sacral arteries extend to the entire upper gluteal region as far as the greater trochanter and coccyx,[191] while the corresponding cutaneous arteries reach only a narrow border region of the gluteal skin by the iliac crest.

The cutaneous nerves and arteries of the medial thoracic back region also show similarities in arrangement and orientation. In both cases the dorsal rami divide into medial and lateral branches, the former perforating the superficial muscular layer immediately adjacent to the spinous processes, and the latter emerging inferiorly between the longissimus dorsi and ileocostalis muscles. In the mutual terminal expansion of the medial and lateral branches, the arteries and nerves also exhibit similar relationships. In both cases the medial branches decrease in size from superior to inferior, whereas the lateral branches correspondingly increase.

Finally they also correspond for the most part in their main direction and in the deviation of this direction from the vertebral segments.

The similarity between cutaneous nerves and arteries is also great in the lateral trunk region. The cutaneous arteries of this region emerge from the intercostal arteries as perforating branches. In corresponding fashion, penetrating cutaneous branches of the intercostal nerves supply this same territory. In describing the cutaneous arteries of this region, a lateral and an anterior group were distinguished.[192] The lateral group was noteworthy in that its points of exit were reliably found at the tip of a slip of origin of the external oblique muscle and in addition by the fact that, on the surface of the lateral chest wall or directly beneath, its cutaneous arteries divide into a branch moving anteriorly and a branch moving posteriorly. The cutaneous nerves of this region are also divided into a lateral and an anterior group, and for the former (perforating lateral branches of the intercostal nerves) an almost identical arrangement is described as for the corresponding cutaneous arterial group. The points of exit are the same. And the division into two cutaneous branches coursing in opposite directions also occurs in the nerves in a completely analogous fashion

190. Territory of the superficial circumflex scapular artery from the subscapular artery.
191. Territory of the cutaneous branches from the gluteal artery.
192. Compare Chapter I.

(anterior cutaneous rami and posterior cutaneous rami). Like the cutaneous arteries, the posterior cutaneous rami spread to the region of skin overlying the lateral edge of the latissimus dorsi muscle. But the region of expansion of the anterior cutaneous rami is considerably larger than that of the corresponding cutaneous arteries.

In the thoracic portions of the region, the cutaneous nerves spread to the skin of the anterior breast as far as the nipple. In the abdominal region they extend as far as the lateral boundary of the rectus abdominis muscle, and the last nerve even sends out descending branches as far as the inguinal region. The cutaneous arteries associated with the anterior group of the region correspond partially, in the lateral thoracic and abdominal territories, according to their orientation and course, to these anterior cutaneous rami of the lateral cutaneous nerves. But even there they do not advance as far toward the median (see Chapter I).

Cutaneous arteries and nerves also correspond in the anterior chest region. However, their origins are not readily comparable.

The most important cutaneous arteries originate from the internal mammary artery, while the cutaneous nerves are the terminal branches of the internal rami of the intercostal nerves. But the direction and territory of expansion of the cutaneous branches are generally similar in both. Like the perforating rami of the internal mammary arteries, the anterior cutaneous nerves of the breast (anterior cutaneous pectoral nerves) perforate the pectoralis major muscle immediately adjacent to the lateral edge of the sternum. After giving off branches to the skin overlying the anterior surface of the sternum, they arborize in a lateralward direction. Branches of the second to fourth anterior cutaneous nerves extend to the skin of the mammary gland.

In contrast, no similarity between cutaneous arteries and nerves is seen in the abdominal region. The growth circumstances of the anterior abdominal wall, which, in the case of the arteries, produce the arrangement of the superficial epigastric arteries and the abdominal branches of the external pudendal arteries, are followed by the nerves in an entirely different manner. The nerves descend diagonally to the median line, ultimately coursing far inferior to their affiliated thoracic vertebral segments.

The cutaneous arteries and nerves in the region of the ischiorectal fossa again exhibit considerable similarity of arrangement.

The pudendal nerve, which supplies this region of skin with its terminal branches, resembles the corresponding internal pudendal artery to a significant degree. Both enter the ischiorectal fossa through the lesser sciatic notch and divide there into their terminal branches.

Even before its exit from the pelvic cavity, the pudendal nerve gives rise to the perforating sacrotuberous ligament nerve[193] from which cutaneous branches pass around the inferior edge of the gluteus maximus muscle to supply

---

193. Schwalbe, p. 981.

the border region of the skin overlying this muscle. A very similar behavior was noted for the internal pudendal artery in Chapter I.

The external hemorrhoidal nerve, which arborizes in the skin overlying the ischiorectal fossa and the external anal sphincter, corresponds very closely to the external hemorrhoidal arteries. The posterior (lateral) scrotal arteries exist in the same sort of analogous arrangement and terminal expansion with the medial perineal nerves and the lateral perineal nerve,[194] and the same correspondence occurs between the penile artery and nerve. The same also holds true for the anterior scrotal nerves, which arise from the ilioinguinal nerve, and for the anterior scrotal arteries from the external pudendal arteries. On those frequent cases where the anterior scrotal nerves originate from the ilioinguinal nerve, the correspondence is even greater. For this nerve, according to its exit point from the femoral fascia and its expansion in the upper femoral region, essentially corresponds to the external pudendal arteries.

In the head as well, clearly recognizable correlations exist between cutaneous arteries and nerves; this is particularly valid for some of their territories of terminal expansion.

The cutaneous expansion of the upper cervical nerve in the occiput, for example, corresponds closely with the territory of the occipital and posterior auricular arteries.

Especially interesting is a comparison of the cutaneous arteries and nerves of the external ear. In Chapter III it was demonstrated that different parts of the external ear received their cutaneous arterial supply from different sources, depending on their embryonic origins from the anterior or posterior edge of the branchial cleft. The cutaneous nerves of the external ear are distributed in a similar fashion. The anterior auricular nerves, which originate from the auriculotemporal nerve, "supply the skin of the anterior edge of the pinna from the superior tip to the tragal notch and the tragus."[195] Their territories of expansion therefore coincide with those of the anterior auricular arteries from the superficial temporal artery.

The main cutaneous territory of the posterior auricular artery on the external ear is innervated by cutaneous branches of the great auricular nerve (from the cervical plexus). The great auricular nerve divides into an anterior and a posterior branch inferior to the earlobe. "One part of the anterior branch spreads out over the medial surface into the skin of the earlobe and reaches the skin of the concave surface of the concha by perforating the auricular cartilage with fine filaments. The posterior branch spreads out over the medial side of the earlobe and in the skin posterior and superior to the ear."[196]

---

194. The ramification of individual cutaneous branches from the lateral perineal nerve onto a small portion of the medial thigh surface also finds its analogs among the cutaneous arteries. A major cutaneous branch from the laterally directed posterior scrotal artery often passes over to the posteromedial thigh area.

195. Schwalbe, p. 842.

196. Schwalbe, p. 905.

Finally, the cutaneous territory of the first branch of the trigeminal nerve coincides almost congruently with the cutaneous territory of the ophthalmic artery from the internal carotid artery. But the cutaneous nerves in this region supply a decidedly larger part of the vault of the scalp than the arteries do.[197]

For the most part, the following individual cutaneous nerves and arterial trunks can also be directly compared with one another according to their origins, orientation, and terminal expansion:

Supratrochlear artery—supratrochlear nerve
Supraorbital artery—supraorbital nerve
Nasal artery—cutaneous branches of the nasociliary nerve

In addition to this congruence of regions of expansion, the far-reaching correspondence in the main orientation of the cutaneous arteries and nerves in the entire head region is noteworthy.[198]

The correspondence in orientation of the cutaneous arteries and nerves, which we frequently found to be very clearly expressed in the trunk and head, leads to an interesting question, which places the significance of a detailed knowledge of the cutaneous arteries in a new light. Therefore let us be allowed to point out this question briefly at the conclusion of this treatise. Some cutaneous diseases show a great regularity in their distribution and in the direction of their spread. It is alleged that diseases like *Herpes Zoster* infection, psoriasis, and others "frequently follow the direction of the cutaneous nerves completely in their appearance and their expansion."[199]

As just one example, it can be mentioned that in these diseases, the involvement of the skin in the lateral thoracic region is arranged in parallel rows which correspond to the ribs in their orientation. Kaposi attributes these circumstances of arrangement primarily to the directions of cleavage lines of the skin.

Perhaps, however, they are actually traceable not so much to the direction of lines of cleavage, but rather much more to the arrangement of the cutaneous arteries which the lines of cleavage have dictated. Even Kaposi concedes that some things remained unexplained by lines of cleavage alone, and he expresses the view that much etiologic information is to be found in the little known fine vascular distribution.

A detailed investigation of this question would certainly lead to interesting results.

197. According to Heiberg (Atlas der Hautnervengebiete), they reach as far as the lambdoidal suture with their terminal branches.
198. Compare the figure of the cutaneous nerves of the head in Schwalbe's Neurologie XX with Figure 8 (Plate V) of this treatise.
199. Kaposi, Pathologie und Therapie der Hautkrankheiten, p. 74+.

Bibliography

1. Arnold, Handbuch der Anatomie des Menschen. II.
2. Gegenbaur, Lehrbuch der Anatomie des Menschen. 1885.
3. Henle, Handbuch der Gefässlehre des Menschen. 1876.
4. Heiberg, Atlas der Hautnervengebiete.
5. His, Anatomie menschlicher Embryonen.
6. His, Ueber den Sinus praecervicalis und über die Thymusanlage.
7. His, Die Retromandibularbucht.
8. Hyrtl, Corrosionsanatomie.
9. Kaposi, Pathologie und Therapie der Hautkrankheiten.
10. Kölliker, Entwicklungsgeschichte des Menschen und der höheren Wirbelthiere.
11. Krause, C. Fr. Th., Handbuch der menschlichen Anatomie. I. 2. 1842.
12. Krause, W., Handbuch der menschlichen Anatomie. 1879.
13. Langer, Ueber die Spaltbarkeit der Cutis.
14. Langer, Die Spannung der Cutis.
15. Rabl, Ueber das Gebiet des Nervus fascialis.
16. Rabl, Zur Bildungsgeschichte des Halses.
17. Sappey, Traite d'anatomie descriptive. II. 1876.
18. Schwalbe, Lehrbuch der Neurologie.
19. Stahel, Anatomie und Chirurgie der Arteria subclavia.
20. Theile, Lehre von den Muskeln und Gefässen des menschlichen Körpers. 1841.
21. Tiedemann, Tabulae angiologicae.
22. Türstig, Mittheilungen über die Entwicklung der primitiven Aorten.
23. Türstig, Untersuchungen über die Entwicklung der primitiven Aorten.
24. Voigt, Chr. A., Beiträge zur Dermato-Neurologie. 1864.